# Agricultural Botany

## N. T. Gill and K. C. Vear

Third Edition, revised by
K. C. Vear and D. J. Barnard

## 1. DICOTYLEDONOUS CROPS

DUCKWORTH

First published in 1980 by
Gerald Duckworth & Co. Ltd
The Old Piano Factory
43 Gloucester Crescent, London NW1

© 1980 by N. T. Gill, K. C. Vear and D. J. Barnard

Cloth ISBN 0 7156 1250 6
Paper ISBN 0 7156 1191 7

British Library Cataloguing in Publication Data

Gill, Norman Thorpe
  Agricultural botany.—3rd ed. revised.
  1: Dicotyledonous crops
  1. Field crops
  I. Title    II.  Vear, Kenneth Charles
  III. Barnard, D.J.
  631      SB 185

  ISBN 0-7156-1250-6
  ISBN 0-7156-1191-7 Pbk

Made and printed in Great Britain by
The Garden City Press Limited
Letchworth, Hertfordshire SG6 1JS

# CONTENTS

# PREFACE

The changes which have taken place during the twenty years since the original publication of *Agricultural Botany* have made extensive revision desirable. Minor revisions, connected mainly with changing use of particular varieties, were made in the second edition and revised reprint, but the time now appears ripe for more thorough rewriting.

Dr Gill, to whose initiative the original work was due, retired in 1969 and did not feel able to take part in this revision, which has been undertaken by K. C. Vear and his former colleague in the Botany Department at Seale Hayne College, D. J. Barnard. While the responsibility for the present edition is theirs, the work as a whole nevertheless still owes much to Dr Gill's ability as a teacher and writer.

The changes involved are not merely those due to advances in knowledge and differences in interpretation, and to the continued work of plant breeders resulting in constant alteration in the cultivars of the commonly grown crops, but also to marked changes in agricultural practice and in particular to the breaking down of the distinction between agricultural and horticultural crops, to the extensive changes consequent on the adherence of Britain to the European Economic Community and to the introduction of metrication. Changes in the organization of agricultural education, and the disappearance of the National Diploma in Agriculture, to which the original edition was in some sense geared, have also made alterations desirable. Since a course in agricultural botany does not always follow previous training in pure botany, it has been thought worth while to include a brief sketch of the general structure, function and classification of plants as an introduction.

Changes in book production methods and costs have made large single-volume textbooks less practicable and this new edition is therefore presented as a series of separate volumes which it is hoped will be more convenient for students. Volume 1 deals with the dicotyledonous crop plants, covering the ground of chapters 8–13 of the original edition; Volume 2 the monocotyledonous crops of chapters 14–17. It is intended that Volume 3 will deal with plant breeding and crop testing, equivalent to chapters 2–7 of the original; Volume 4 with weeds, chapters 18–20; Volume 5 with plant diseases, chapters 21–29; and Volume 6 with crop physiology, a section of agricultural botany not covered in previous editions.

Thanks are due to all those who assisted in the preparation of earlier editions, and for this new edition particularly to Dr P. S. Wellington, Director of the National Institute of Agricultural Botany, for renewing permission to make use of the publications of the N.I.A.B., invaluable to any agricultural botanist in Britain, and for advice and assistance on the difficult problem of the infra-specific classification of crop plants. A debt is also due to Professor A. H. Bunting and Dr J. K. Jones for valuable criticism of the earlier editions, to colleagues at Seale Hayne for helpful discussion and advice over the years, to Dr F. A. Vear for reading and assisting with some sections, to the publishers for their skilled advice and help, and, not least, to a long series of students from whom we have learnt much.

K.C.V.
D.J.B.

# 1

# PLANTS IN GENERAL

Agricultural botany is the study of the plants which are of importance in farming; that is, of the plants which are used as crops and also of those which, by occurring as weeds or by causing disease, are hindrances to crop growth. It is thus a specialized study of a limited number of types of plants. A preliminary study of the general structure and functions of plants is therefore desirable.

## TYPES OF PLANTS

Plants, like animals, are living organisms. They differ from animals in that, for the most part, they are self-supporting (holophytic); that is, they are not dependent on pre-existing organic matter. They are mainly non-motile, and their cells, the fundamental units of living matter of which they are composed, are typically surrounded by a relatively rigid cell wall. Like animals they show some form of sexual reproduction, which allows recombination of the genetic characters of two individuals. As a result of this, the progeny may differ from the parents, and consequently evolutionary change may take place.

The plant kingdom, with something approaching half a million species of very widely varying plants, has been classified by different authorities in different ways, and a large number of subdivisions have been used. Since many of these are of little or no agricultural importance, a much simpler classification is adequate for the present purpose, and the plant kingdom can be divided into the following very broad groups:

**Thallophytes**   Algae, Fungi, Bacteria
A very heterogeneous group including many different and probably unrelated series of plants, agreeing only in the negative character of having separate or uniseriate cells not forming true tissues; usually split into a large number of smaller groups.

**Bryophytes**   Mosses, Liverworts
A fairly uniform but probably rather isolated group of plants; true

tissues produced by cell-division in more than one plane, but specialized conducting tissue absent.

**Pteridophytes**   Ferns, Clubmosses, Horsetails
Flowerless, but structurally akin to the flowering plants and gymnosperms, and having the same type of specialized conducting tissue, and therefore sometimes grouped with them to form the division *Tracheophyta*.

**Spermatophytes**   Seed-bearing plants, divisible into:
  **Gymnosperms**   Pines and similar plants in which the seeds are produced on the outside of leaf-like scales.
  **Angiosperms**   Flowering plants in which the seeds are borne inside a closed fruit.

The ferns, gymnosperms and angiosperms are probably closely related and form a genuine evolutionary series; they are sometimes treated as together forming the *Pteropsida*, a major group within the *Tracheophyta*.

If we except the fungi, which are an unusual type of plant unable to synthesize organic matter from inorganic, and therefore either saprophytic or parasitic, and also the bacteria, which are also mainly saprophytic and have a very different structure from that of most plants, we may consider these plant groups as forming a series of steps, of increasing complexity and of increasing adaptation to life on dry land. In plants sexual reproduction is associated with alternation of generations, so that the *gametophyte* generation, which produces gametes, the units which fuse in sexual reproduction, is succeeded by a *sporophyte* generation, which produces spores which give rise to the gametophyte again. In the cells of the gametophyte the *chromosomes*, the structures which carry the genetic information which determines the plant's characters, are present in the *haploid* number. When the two gametes fuse to form the zygote, this number is doubled to the *diploid* condition characteristic of the sporophyte generation. The nuclei in the cells of the sporophyte will thus contain two sets of chromosomes, one from each gamete. Each chromosome in one set will normally have a corresponding (homologous) chromosome (of similar size and shape, but not necessarily carrying identical genetic information) in the other set. In both generations increase in size takes place as a result of cell division by a type of fission known as *mitosis*, in which each chromosome is exactly reproduced in each of the two new cells, so that there is no change in either the number of chromosomes or in the genetic information carried in the cells. When, however, the sporophyte comes to produce spores, a special type of reduction division known as *meiosis* takes place, in which half the chromosomes (one from each homologous pair) pass to one new cell and half to the other, so that the spores, and consequently the gametes to which they eventually give rise, have only the haploid number of chromosomes and may differ in the genetic information they carry. Genetic variation is often increased by exchange of material between homologous chromosomes at meiosis.

In most plants, as in animals, the gametes are distinguishable into small and usually motile male gametes, and larger non-motile female gametes. In the more primitive groups the gametophyte is hermaphrodite and produces both male and female gametes; the male gametes are free-swimming and the presence of external water is therefore essential. In the bryophytes the gametophyte is the more conspicuous and longer-lived generation, but in the higher groups the position is reversed. In the pteridophytes it is the sporophyte which is the conspicuous fern plant, and the gametophyte is a small short-lived independent structure, the prothallus. The prothallus of the true ferns is hermaphrodite, but in the more advanced pteridophytes *heterospory* is found; large *megaspores* give rise to female, and small *microspores* to male gametophytes.

In the gymnosperms and angiosperms megaspores produce small female gametophytes in *ovules* borne on the sporophyte. Microspores, producing a male gametophyte reduced to one or two nuclei, are shed as *pollen grains*, which are carried by wind or other agents (pollination) to enable the male gametes, which are motile only in the more primitive gymnosperms, to reach the female gametophyte. In the higher plants therefore, external water is no longer essential for reproduction. The zygote develops inside the ovule into a young sporophyte plant, the *embryo*. This is provided with a reserve food store, which may be either internal or external, and becomes dormant. The dormant embryo with its food reserve and protected by the *testa*, derived from the integuments of the ovule, is shed as the *seed*. Only under favourable conditions does the embryo resume growth and germination take place.

In the gymnosperms the ovule (*megasporangium*) is borne on the surface of a leaf-like cone scale (*megasporophyll*). In the angiosperms further protection is given by the ovule being borne inside a closed *ovary* formed from one or more *carpels* (megasporophylls). A receptive area at the top of the carpels forms the *stigma*, and pollen grains brought into contact with this germinate to produce a pollen tube by means of which the two male gametes, consisting of nuclei only, reach the *embryo-sac* (female gametophyte) in the ovule. The female gametophyte consists usually of eight haploid nuclei; one male nucleus fuses with one of these to give the embryo; the other male nucleus fuses with two others to form the triploid *endosperm*, the food reserve tissue which may either be absorbed by the embryo during its development or remain as a separate structure which will not be drawn upon until the embryo recommences growth when the seed germinates.

MACROSCOPIC STRUCTURE OF DICOTYLEDONS

The embryo in the angiosperms consists normally of an axis bearing either one or two lateral structures, the *cotyledons*. This difference provides the essential basis of the classification of the angiosperms into two sub-classes, monocotyledons and dicotyledons. The lower part of this axis is the *radicle*, which develops into the primary root. The root increases in length by division of cells near its tip; it branches by the production of lateral roots which arise some distance behind the root-tip. The root is usually positively geotropic and grows down into the soil. It normally bears no organs other than roots. Above the root the axis continues as a short transition region, the *hypocotyl*, to the *cotyledonary node*, where the two cotyledons (in dicotyledons) are attached. Above this is the *plumule*, usually very short in the embryo, which will develop into the shoot, consisting of stem (epicotyl) and leaves. The leaves are borne at the nodes of the stem, and at these nodes buds are produced in the axils of the leaves; these may grow out into axillary branches. The growing point of the shoot is at its tip, where new leaves are also initiated.

Leaves are essentially organs in which a large number of physiological processes are carried out. These include *photosynthesis*, the essential process resulting in the formation of organic compounds from carbon dioxide and water. Energy is necessary for this, and is obtained by intercepting the radiant energy from the sun. The leaf is therefore typically a thin plate of tissue, which allows a relatively small amount of material to provide a large intercepting area. The process of photosynthesis cannot be completely efficient, that is more energy must always be intercepted than can be utilized in the production of the organic compounds; the excess energy, which would otherwise appear as heat and raise the leaf temperature to destructive levels, is used up in *transpiration*, the evaporation of water from the leaf.

The stem supports the leaves in a position in which they can carry out photosynthesis and transpiration. Conducting strands in the leaf are continuous with those in the stem and root and the evaporation of water from the leaves results in water passing up the stems from the

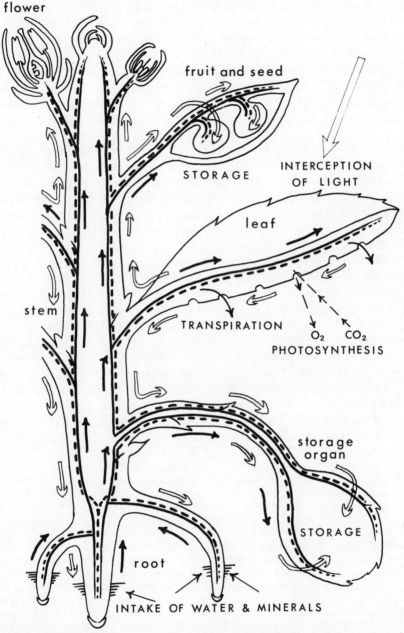

flower

fruit and seed

STORAGE

INTERCEPTION
OF LIGHT

leaf

stem

TRANSPIRATION

O₂   CO₂

PHOTOSYNTHESIS

storage
organ

STORAGE

root

INTAKE OF WATER & MINERALS

Fig. 1. Diagram of a generalized flowering plant in longitudinal section to show the functions of the various parts. Solid lines and arrows, xylem and transport of water and minerals; broken lines and open arrows, phloem and transport of elaborated materials.

roots, which absorb it from the soil. Mineral ions necessary for the life of the plant are also absorbed by the roots and are carried in this *transpiration stream*. Organic compounds are synthesized in the leaves and from there *translocated* in the stem and root to all parts of the plant. If these materials are in excess of the immediate needs of the plant, and storage is necessary, they are stored in some special *storage organ*, which may be a swollen part of the root or stem or leaf base. In *annuals*, plants which live for one year only, there is usually little storage except that of the seed reserves produced after flowering; in *biennials*, living for two years, there is sometimes a substantial storage of reserve material at the end of the first year, used up in flowering and seed formation in the second year. In some *perennials* the above-ground parts may die off in winter, or during summer drought, and the plants persist only as dormant underground storage organs, which recommence growth when conditions become favourable.

Eventually, often in response to an external stimulus such as the appropriate day length, the plant produces *flowers*. These may be considered as highly specialized short shoots. The short axis of the flower (*receptacle*) bears, spirally in the more primitive types, in whorls in the more evolved, a number of structures regarded as modified leaves. The outer ones, *sepals*, collectively forming the *calyx*, are usually green and protect the inner structures in the unopened flower bud. Inside these are the *petals*, collectively forming the *corolla*, often brightly coloured and conspicuous, and serving to attract insects where these act as pollinating agents. In some flowers, especially those which are wind pollinated, all the outer structures are similar and often greenish, and are then spoken of as *perianth segments* or *tepals*, collectively forming the perianth.

---

Fig. 2. Diagram of the life-history of a generalized flowering plant. A, plant in flower. B, single flower in half-section. C, single anther at pollen-mother-cell stage, in transverse section. D, mature anther dehiscing and shedding pollen grains. E, gynaecium in vertical section, with single ovule at megaspore-mother-cell stage. F, the same after pollination, with eight haploid nuclei in embryo-sac and two male nuclei in pollen tube. G, ovule only shown at slightly later stage; the two male nuclei have entered the embryo-sac, where the three functional female nuclei remain, the others degenerating. H, the same after fertilization, with one diploid nucleus which develops to form the embryo, and one triploid nucleus which gives rise to the endosperm. I, developing fruit with single seed consisting of embryo embedded in endosperm and surrounded by testa formed from integument of ovule. J, seed released from fruit and germinating. K, seedling. *c*, corolla; *cl*, carpel; *ct*, cotyledon; *e*, embryo; *en*, endosperm; *es*, embryo-sac; *i*, integument; *k*, calyx; *m*, micropyle; *mn*, male nuclei; *n*, nucellus; *o*, ovule; *ov*, ovary; *p*, pericarp; *pg*, pollen grain (shown disproportionately large); *pl*, plumule; *pt*, pollen tube; *r*, radicle; *s*, style; *sg*, stigma; *t*, testa.

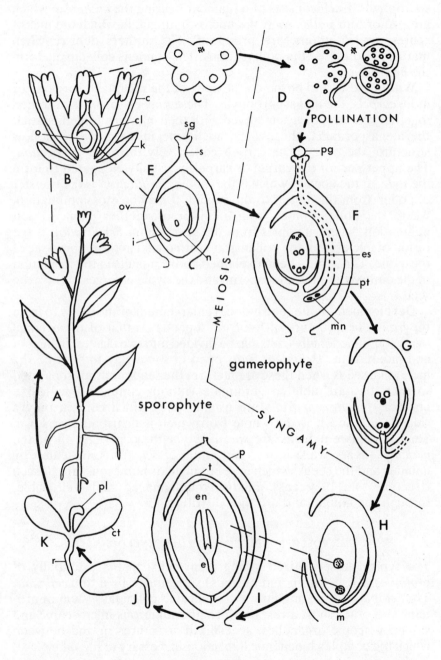

Within the corolla (or perianth) are the *stamens*, which are micro-sporophylls. Each consists of a filament bearing the *anther*, in which are two or four *pollen sacs*, the microsporangia, in which the micro-spores, pollen grains, are produced. The anthers dehisce when mature, setting free the pollen grains. The stamens collectively form the *androecium*.

Within these, and borne on the apex of the receptacle, are one or more *carpels*, the megasporophylls. These may be separate or joined together, but are always incurved with their margins joined, so that the megasporangia, the ovules, are borne inside a closed hollow structure, the ovary. The carpels collectively form the *gynaecium*. The upper part of the carpel or carpels is usually elongated to form the *style*, at the apex of which is the *stigma*. *Pollination* is the transfer of pollen from the anther to the stigma; it may be cross pollination, where the pollen comes from a flower of another plant, or self pollination, where it comes from the same plant. *Fertilization* is the fusion of nuclei from the pollen grain with nuclei in the embryo sac of the ovule. This, as has been explained above, results in the formation of the embryo and the endosperm, and the ovule enlarges to form the seed.

Development of the seed induces enlargement of the ovary to form the *fruit*. The structure of the fruit depends on that of the original ovary, but functionally fruits can be divided into two classes, dehiscent and indehiscent. *Dehiscent fruits* open in some way to set free the individual seeds when these are mature; the seeds are thus separated so that they are able to germinate without competing with one another. *Indehiscent fruits* are usually ones which contain only a single seed, and it is the whole fruit which is distributed or sown. Some indehiscent fruits are *succulent*; in these part or all of the *pericarp* (ovary wall) is soft and edible. Such fruits are eaten by animals and the seeds, which may be single or numerous, are of such structure that they pass, relatively undamaged and still viable, through the animal and are thus distributed.

### MICROSCOPIC STRUCTURE OF DICOTYLEDONS

The typical unspecialized young plant cell consists essentially of *protoplasm* (cytoplasm and nucleus) surrounded by a thin *cell wall*. The *cytoplasm* is an immensely complicated colloidal system of pro-teins and other substances, in which are numerous microscopic and sub-microscopic organelles, specialized structures in and between which the complex biochemical reactions necessary to life take place. In the nucleus are the *chromosomes*, specialized structures containing nucleic acids which embody the genetic information passed on to the plant from its parents. This information determines the structure and behaviour of the cell and of the whole plant. Adjacent cells are separated by a thin *middle lamella* composed of pectic substances. A thin *primary wall*, in contact with the middle lamella, is secreted by the cytoplasm of each cell. This wall contains *cellulose*, a long-chain carbohydrate which is insoluble but permeable to water. *Plasmodesmata*, fine threads of cytoplasm, pass through the wall to connect adjacent protoplasts.

These unspecialized young cells are capable of further division; they are found in the growing points of the plant and in other regions where cell multiplication is taking place. Such regions are known as *meristems*, and the cells thus as meristematic cells, collectively forming *meristematic tissue*. Meristematic tissue forms satisfactory food for all types of animals, since it consists very largely of protoplasm, similar in general composition to that of the animal. However it never forms more than a very small proportion of the plant, and therefore is of very minor importance in the selection of food crops. Even where, as in mustard and cress, the plants are eaten as young seedlings, many of the cells have already become differentiated, and are no longer meristematic.

The majority of the cells produced by cell division in the growing points of the plant develop into *parenchyma* cells. They increase in size and become vacuolated, and the cell wall increases in thickness and in cellulose content; *intercellular air spaces* also develop. These changes mean that typical parenchyma, although edible, has a high water content and a high cellulose content. It forms a very satisfactory food for ruminant animals which have an alimentary canal adapted to the intake of a large volume of food, and capable of digesting cellulose. For human beings with a relatively small alimentary canal, unable to digest cellulose, typical parenchyma is not a sufficiently concentrated food to form a staple diet. Most green vegetables and fruits consist largely of this typical parenchyma, but they normally form only a small part of the human diet, and although they do contribute to the intake of carbohydrate and protein, are usually eaten mainly for their mineral and vitamin content, or to provide the necessary bulk in an otherwise more concentrated diet.

Parenchyma can also act as a storage reservoir for carbohydrates and other elaborated compounds; where special storage organs are developed these consist largely of parenchyma. The cells of this storage parenchyma may contain large quantities of starch or sugar, so that the percentage dry matter is higher (up to say 25%) and the ratio of readily digested organic matter to cell wall material much higher than in typical parenchyma. Vegetative storage organs consisting largely of such parenchyma therefore provide more concentrated food, and are used not only as root-crops for animal consumption, but also as root and stem vegetables which can form an important part of human diet. In many cases such vegetative storage organs have a period of dormancy before growth recommences, and their nutrients are used up in the production of further tissue. It is therefore possible for them to be harvested and stored for some months before being eaten. Plant structures like leaves and young unswollen stems which consist mainly of typical low dry-matter parenchyma are not normally storable unless some special method of preservation such as freezing or pickling is used.

In seeds an even more concentrated form of storage parenchyma is found. This is the parenchyma of the endosperm or of the cotyledons, where the dry matter in the ripe seed may reach 90%, and where a very large proportion of the dry matter consists of readily digestible starch, oil or protein. Seed reserves thus provide not only the most concentrated type of vegetable food, suitable for both ruminants and non-ruminants, but also the most convenient type, in that the seed remains dormant indefinitely if it is kept dry, and can therefore be stored for some years, if necessary, without deteriorating.

These three types of parenchyma, varying in their dry-matter content, but agreeing in consisting of material which is wholly digestible by ruminants and, apart from the relatively thin cellulose cell wall, by non-ruminants as well, are the essential elements in plants as food for animals. Other types of differentiated cells are of little value as animal food, and the plants utilized as food, or cultivated as food-crops whether for human beings or for domesticated animals are essentially those in which there is a high proportion of parenchyma tissue, either in the whole plant, or in that part of it which is utilized.

No plant can consist solely of parenchyma, since although all the essential biochemical reactions can occur in parenchyma cells, the rate of transport through parenchyma tissue is very slow, and such tissue is mechanically weak. In all higher plants some cells differentiate, not as parenchyma but as specialized conducting cells or as elements to give mechanical strength. Two largely separate conducting systems are involved; one to transport water and dissolved minerals from the root to all other parts of the plant and particularly to the leaves, and the other to transport elaborated organic compounds from the leaves to the other parts of the plant.

The transport of water and dissolved minerals takes place in cells which become much elongated in the direction of transport; they thus differ from the usually more or less isodiametric parenchyma cells. In

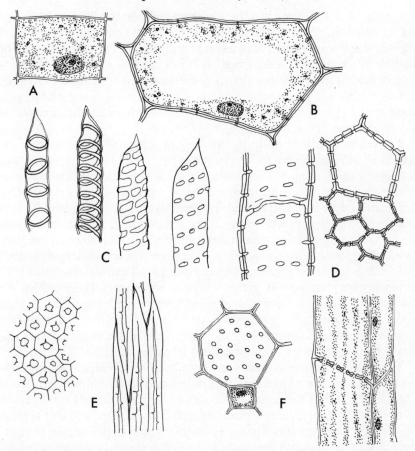

Fig. 3. Types of cells, diagrammatic. A, meristematic cell. B, parenchyma cell. C, parts of four types of tracheid; annular, spiral, scalariform and pitted. D, part of vessel in longitudinal section and transverse section of part of xylem showing vessel and tracheids. E, sclerenchyma in transverse and longitudinal section. F, phloem; transverse section of sieve tube through sieve plate and adjoining companion cell, and the same in longitudinal section.

addition to a change of shape as the cells differentiate there is also a change in cell wall composition. The energy for water transport is provided by radiant energy intercepted by the leaves evaporating water from the free surfaces of parenchyma cells into the intercellular spaces. The water vapour diffuses through these spaces into the external atmosphere via the stomata, which are openings in the epidermis, the outer layer of cells covering the leaf. The loss of water from the parenchyma cells results in water being drawn from adjacent cells and thus water is drawn up through the stem under tension.

Since the water columns in the elongated water-conducting cells are under tension, the flow (transpiration stream) can only proceed if their cell walls are rigid enough not to collapse. A cellulose wall is not sufficiently rigid, and accordingly is reinforced by a lignified thickening, in which the cellulose fibrils are embedded in a rigid matrix of *lignin*, which has greater mechanical stiffness. Such elongated water-conducting cells with walls wholly or partially lignified are known as *tracheids*; the tissue they form is *xylem*.

Cells which differentiate in this way in a part of the stem or root which is still increasing in length must be capable of extension, and the lignin thickening is laid down either as separate hoops (*annular* tracheids) or as a continuous spiral band with the turns separated by unlignified wall (*spiral* tracheids); these form the *protoxylem*. Xylem cells which differentiate later, after extension is complete (*meta-xylem*), have lignin disposed in longitudinal as well as transverse bars (*scalariform* tracheids) or have an almost continuous thickening with only small separate areas of unthickened wall (*pitted* tracheids). In angiosperms, but not in gymnosperms, some very large xylem elements are formed in which the end walls are resorbed, so that contiguous elements come to form a continuous lignified tube, known as a *vessel*. Since lignin is largely impermeable the cell contents die when the tracheid or vessel-element becomes mature.

Typical xylem thus consists essentially of dead cells, without contents, and with the walls heavily impregnated with lignin, which is not digested by either ruminant or non-ruminant animals. It is therefore of no feeding value; woody plants which consist largely of this typical lignified xylem may be of great economic value in producing timber, but cannot be used as food, although of course they may produce largely unlignified leaves or fruits which may be edible.

Mechanical strengthening tissue, *sclerenchyma*, also develops thickened lignified walls, and the mature cells, which are often considerably elongated, with the walls so much thickened that only a small central lumen is left, are also dead and without contents. Sclerenchyma is thus, like lignified xylem, of no value as food, although it may sometimes be of economic value in providing fibres for use in textiles or cordage. Some unlignified mechanical tissue exists; cellulose-walled fibres occur for example in the stem of flax. *Collenchyma* is a further type of strengthening tissue; in this still-living cells have their walls much thickened by cellulose, particularly at the angles.

Transport of elaborated materials takes place in the *phloem* tissue, in special elongated cellulose-walled cells, which are living but enucleate, and with the end walls perforated so that strands of cytoplasm are continuous from cell to cell. These perforated cross walls are known as *sieve plates*, and the cells of which they form part as *sieve tube elements*; each sieve tube element is accompanied by a smaller nucleate cell known as a *companion cell*. Phloem tissue composed of sieve tubes and companion cells, or with associated parenchyma, is edible, and its cell contents may be of high nutritive value, but phloem is frequently interspersed with or adjoined by fibres, and it is rarely that a large volume of edible phloem is formed.

### The primary structure of the stem

The apical meristem of the stem may be distinguished into an outer *tunica*, giving rise to the external layers, and an inner *corpus*. Leaves originate as raised areas formed from the tunica shortly behind the domed apex. Cells within the outer layer (*epidermis*) differentiate as the parenchyma of the *cortex*, and those in the centre as the parenchyma of the *pith*. Between these a series of separated groups of cells elongate but retain their undifferentiated structure to form, as seen in transverse section, a ring of *procambial strands*. These procambial strands develop into *vascular bundles*, the cells nearest to the pith

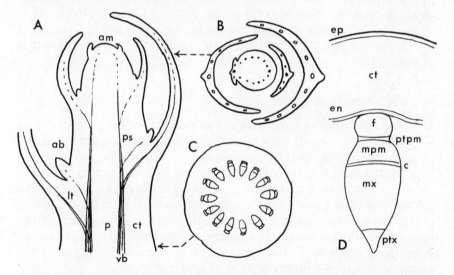

Fig. 4. Primary structure of dicotyledonous stem. A, longitudinal section of upper part of shoot. B, C, transverse sections at two levels. D, enlarged view of single vascular bundle of C and adjacent tissues. *ab,* axillary bud; *am,* apical meristem; *c,* cambium; *ct,* cortex; *en,* endodermis; *ep,* epidermis; *f,* fibres; *lt,* leaf trace; *mpm,* metaphloem; *mx,* metaxylem; *p,* pith; *ps,* procambial strand; *ptpm,* protophloem; *ptx,* protoxylem; *vb,* vascular bundle.

differentiating as protoxylem and those nearest to the cortex as protophloem. In the somewhat older part of the stem, where elongation has ceased, the remaining cells of the procambial strands differentiate as metaxylem and metaphloem. A line of cells at the junction of xylem and phloem often remains meristematic as the *cambium*. At the nodes one or more of the vascular bundles pass into each leaf. The stem structure thus consists at this stage of a central pith surrounded by a ring of vascular bundles and this in turn surrounded by the cortex and epidermis. Parenchyma forming the vascular *rays* (medullary rays) extends from pith to cortex between the bundles; single rings of cells immediately outside the vascular bundles may be distinguishable as *pericycle* and *endodermis*, but are not usually well defined in the stem.

The position at this stage is referred to as the *primary structure* of the stem; in herbaceous stems this may be the final structure, and no development of the cambium takes place. Cells arising from the pericycle immediately adjacent to the protophloem frequently differentiate as sclerenchyma, and form a strand of fibres adjoining the phloem, so that the bundles can be referred to as fibro-vascular bundles. In herbaceous stems these are usually large and well defined, and provide the necessary mechanical strength (sometimes supplemented by collenchyma in the cortex) for a relatively short stem. This arrangement of mechanically strong tissues in a ring near the circumference provides the greatest stiffness for a given amount of material.

### The primary structure of the root

The root apex differs from that of the stem in forming a *root cap*, a shield-like structure of expendable cells constantly renewed as the outer cells are damaged by passage of the root tip through the soil. Behind this meristematic cells differentiate in a way similar to that in the stem, but in such a manner as to give a different arrangement of tissues. Separate procambial strands, surrounded by a small amount of parenchymatous *ground tissue* develop in the central part of the root, those differentiating as xylem alternating with those which develop as phloem. Protoxylem arises at the outer margin of the xylem strands and metaxylem develops centripetally; metaxylem may develop right to the centre, so that a single combined xylem group is formed, or some ground tissue may be found at the centre, and the individual xylem strands remain separate. Roots are referred to as diarch, triarch, tetrarch, etc., according to the number of protoxylem groups present. Alternating with the xylem strands, and on

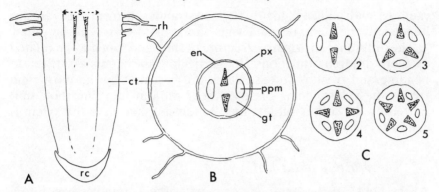

Fig. 5. Primary structure of dicotyledonous root. A, root-tip in longitudinal section. B, young root in transverse section. C, diagrams of stele only of four types of root; 2, diarch; 3, triarch; 4, tetrarch; 5, pentarch. *ct*, cortex; *en*, endodermis; *gt*, ground tissue of stele; *ppm*, primary phloem; *px*, primary xylem; *rc*, root cap; *rh*, root hairs.

different radii, are small phloem strands. Immediately outside the phloem is a ring (as seen in transverse section) of cells forming the pericycle and outside this the endodermis, usually much more conspicuous than in the stem, with the cells thickened on their radial walls by a *casparian strip*.

The structures so far described constitute the *stele*, usually of much smaller diameter than the corresponding part of the stem: this is surrounded by a wide parenchymatous cortex, bounded on the outside by the epidermis. This root epidermis is often referred to as the *piliferous layer*, since some of its cells grow out as *root hairs*. These are usually active only for a short period and then shrivel, and are not usually visible on the older parts of the root more than two or three centimetres from the root apex. Branch roots appear only on these older parts of the root; they originate by division of cells of the pericycle, usually opposite one of the protoxylem groups. The dividing cells develop into the apical meristem of a lateral root, which grows out through the endodermis, cortex and epidermis. The structure of the lateral roots is similar to that of the main root, and their xylem and phloem become connected with those of the main root.

The xylem and phloem strands of the root are continuous with those of the stem; this involves the strands dividing and twisting in the transition region of the hypocotyl, just below the cotyledonary node, so that the few xylem strands of the root, with protoxylem external to metaxylem, continue as the more numerous xylem strands of the epicotyl, with protoxylem internal to metaxylem. Similarly the phloem strands, on different radii in the root, come to lie on the same radii as, and adjacent to, the xylem strands, to form the vascular bundles of the stem, both curving outwards to surround the pith.

The difference in primary structure of stem and root may be related to their different functions. In particular, the stem is subject to bending and twisting forces, which are best met by the most rigid tissue, the xylem, being disposed near the outside. The root, supported by the soil, is subject only to tension, and here the xylem is centrally placed.

### The structure of the leaf

At the nodes of the stem, one or more vascular bundles pass into a leaf, and extend through the *petiole*, where they are surrounded by rather compact parenchyma, to form the central vein, or mid-rib, of the leaf blade (*lamina*). Repeated branches are given off from the main veins, so that the whole leaf blade is traversed by a network of progressively smaller vascular bundles. Thus none of the *chlorenchyma* cells, in which the important biochemical activities of the leaf take place, is far removed from tracheids and sieve-tubes responsible for transport. Chlorenchyma cells are parenchyma cells which contain *chloroplasts*: these are cytoplasmic bodies in which alone is found *chlorophyll*, the group of pigments responsible for absorption and utilization of light energy. In many leaves the chlorenchyma is arranged as an upper *palisade layer*, in which the cells are elongated at right angles to the leaf surface, have numerous chloroplasts, and are separated by relatively small intercellular spaces, and a lower layer of *spongy mesophyll*, with cells more irregular in outline, with fewer chloroplasts, and with extensive intercellular spaces. The whole leaf is covered by the *epidermis*, continuous with that of the stem, and, like that of the stem, pierced at intervals by *stomata*. Each stoma consists of a pair of guard-cells which, with changes in turgor, change the size of the aperture between them, through which the intercellular spaces communicate with the external atmosphere, allowing gas exchange to take place. In a typical dorsiventral leaf, with a well developed palisade layer, and adapted to relatively dry conditions, stomata are usually confined to the lower (abaxial) epidermis. In many food crops, particularly those in which the leaf is edible, there is no well marked distinction into palisade and spongy mesophyll, and stomata are present in the upper (adaxial) epidermis as well as the lower. Such leaves are not adapted to conditions of water stress, where the loss of water by transpiration tends to exceed the supply of water to the leaves from the roots.

Loss of water through the cells of the epidermis is largely prevented

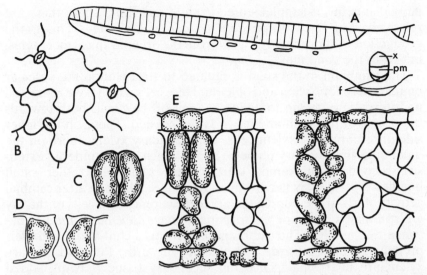

Fig. 6. Leaf structure. A, diagrammatic transverse section of part of dorsiventral leaf. B, surface view of epidermis with stomata. C, D, single stoma in surface view and section. E, part of transverse section of dorsiventral leaf more highly magnified (cell contents shown in some cells only). F, as E, but leaf with uniform mesophyll.

by the *cuticle*, a layer of wax-like cutin which covers their outer walls; true waxes may also be secreted. Transpiration thus takes place largely through the stomata. In some *xerophytic* plants, that is plants adapted to dry conditions, the stomata open at the bottom of depressions in the epidermis, or the leaves are densely hairy, so that the rate of outward movement of water vapour is reduced. In other xerophytes transpiration is not reduced, but the leaves have a high proportion of sclerenchyma, and do not readily wilt when their water content is reduced.

Leaves are usually relatively short-lived structures, in which a period of growth is followed by maturity and this in turn by senescence. In many plants abscission of dead leaves takes place; a special temporary meristem, the *abscission layer*, forms at the base of the petiole and produces a layer of impervious cork cells, along which separation takes place, and which acts as a protective cover to the resultant leaf scar. *Deciduous* plants are ones in which all leaves absciss at one time (autumn in temperate climates) so that transpiration is greatly reduced during the unfavourable period of the year.

### Secondary thickening of the stem

The structures so far described form the primary body of the plant; in most plants this is supplemented in stem and root, where these are relatively long-lived organs, by *secondary thickening*. This is the production of additional cells, not from the original apical meristems, but from the cambium.

The cambium in the stem is situated to start with in the vascular bundles between xylem and phloem. Here cells which have remained undifferentiated begin to divide and cut off new cells both towards the centre of the stem and towards the outside. Those on the inner side differentiate as xylem cells, the secondary xylem, those on the outer side as secondary phloem. Stems which are to undergo extensive secondary thickening usually have numerous rather small bundles separated by narrow medullary rays, and soon after cambial activity starts in the bundle the adjacent parenchyma cells of the ray are stimulated to divide, so forming an *interfascicular cambium* between the bundles. The cambium then forms a continuous cylinder, producing xylem inwards and phloem outwards. Pith and primary xylem are unaffected, since the secondary tissue is being added outside these; but the production of any large amount of secondary xylem necessitates an increase in the circumference of the cambial

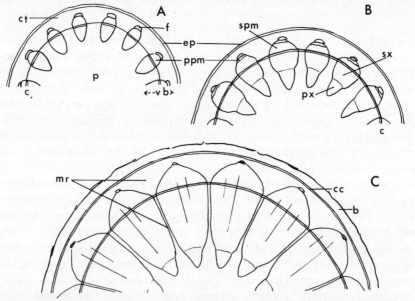

Fig. 7. Secondary thickening of dicotyledonous stem. A, initiation of cambium. B, intermediate stage with some secondary tissue developed. C, later stage with further thickening and development of cork cambium and bark. *b*, bark; *c*, cambium; *cc*, cork cambium; *ct*, cortex; *ep*, epidermis; *f*, fibres; *mr*, medullary ray; *ppm*, primary phloem; *px*, primary xylem; *spm*, secondary phloem; *sx*, secondary xylem; *vb*, vascular bundle.

ring, and in that of all layers external to this. The epidermis and cortex can only respond to the resultant tension by cell division or enlargement to a very limited extent, and normally they are soon replaced by a *phelloderm*. Cells, usually in the cortex, divide and form a new meristem, the *cork-cambium*, a ring of cells behaving in the same way as the true cambium, but producing outwards regularly arranged cuboidal cells without intercellular spaces, and with the walls becoming impregnated with suberin, a wax-like substance impermeable to water. These cells form a *cork layer*, through which gas exchange is possible only via the *lenticels*, small areas in which the cork cells are rounded and loosely arranged. The cork layer together with the external cells which die as a result of its formation form the *bark*. A small number of parenchymatous cells cut off inwards by the cork-cambium form a *secondary cortex*.

If secondary thickening continues, so that a large mass of secondary xylem is produced, the first-formed phelloderm is unable to expand sufficiently, and a further cork-cambium arises inside it, so that the first cork-cambium and its products die and are sloughed off as bark. The process is recurrent, and thus after a time the latest cork-cambium is arising in the secondary phloem, and only the most recently formed phloem cells are active, the older phloem being gradually shed as bark. The bark is smooth where the cork-cambium arises as a continuous ring; fissured and irregular where it arises as a series of short overlapping segments.

Thus a typical old woody stem shows a central pith (the cells of which are by now dead) with the original primary xylem immediately around it, surrounded by a massive cylinder of secondary xylem, of which the inner part forms the *heart-wood*. In this no living cells remain, and the tracheids and vessels are blocked, so that no water conduction takes place in it; the outer sap-wood, consisting of the secondary xylem formed during the last few years, contains some living parenchyma cells and its tracheids and vessels are fully functional. As the secondary xylem increases in thickness provision for transport in a radial direction becomes necessary; some cells produced by the inter-fascicular cambium are elongated radially to form the primary medullary rays, and later other, secondary, rays arise. In trees of temperate regions *annual rings* are usually visible since the cambium ceases activity in winter, and the xylem tissue produced in spring is distinguishable, by larger size of tracheids or by a different frequency of vessels, from that of the previous autumn. The massive xylem cylinder, with its medullary rays and annual rings, is surrounded by the very thin ring of cambium, and this in turn by a narrow zone of active phloem and this again by the cork-cambium and bark.

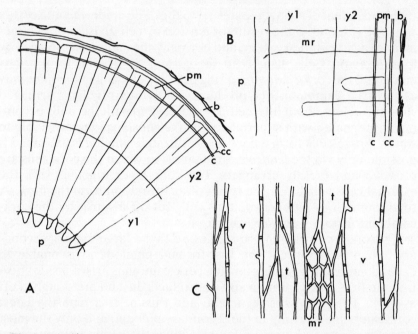

Fig. 8. Two-year-old woody stem. A, in transverse section. B, in radial longitudinal section. C, part of xylem in tangential longitudinal section, more highly magnified. *b*, bark; *c*, cambium; *cc*, cork cambium; *mr*, medullary ray; *p*, pith; *pm*, phloem; *t*, tracheid; *v*, vessel; *y1*, first year xylem; *y2*, second year xylem.

## *Secondary thickening of the root*

Secondary thickening takes place in the root in a similar way to that in the stem, but the origin of the cambium is different. The cambium first appears as short arcs of dividing cells in the ground tissue on the inner side of the phloem groups; these spread around the protoxylem groups to form a continuous line which is at first elliptical or stellate in outline, according to the number of protoxylem groups present. Production of secondary xylem inwards is at first more rapid within the phloem groups, so that the cambium soon becomes circular in outline. Thereafter secondary thickening proceeds exactly as in the stem, the primary rays developing opposite the protoxylem strands. A cork-cambium develops in the pericycle, and the whole of the cortex is sloughed off rapidly leaving only the stele with its developing phelloderm. A root at this stage has a much smaller diameter than a stem of the same age, but with increasing age the difference becomes less marked, and an old woody root may be essentially similar to an old stem except for the absence of the remains of a pith, and the different arrangement of the very small amount of primary xylem.

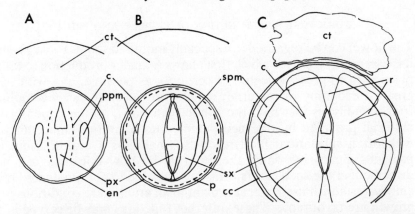

Fig. 9. Secondary thickening of diarch root. A, initiation of cambium. B, development of secondary xylem and phloem. C, later stage with cork cambium developed and cortex sloughing off. *c,* cambium; *cc,* cork cambium; *ct,* cortex; *en,* endodermis; *p,* pericycle; *ppm,* primary phloem; *px,* primary xylem; *r,* rays; *spm,* secondary phloem; *sx,* secondary xylem.

Old secondarily thickened roots have of course no absorptive function, which is carried out solely by the very young roots and root-hairs.

### Secondary thickening in crop plants

Typical woody stems and roots, in which living cellulose-walled cells form only a very small fraction of the total volume, are clearly not suitable for food, and woody plants can only be utilized as food where their leaves (which do not undergo secondary thickening) can be browsed by animals, or where they produce edible fruits or seeds. Secondary thickening is, however, the main method by which plant organs increase in size. There are a few examples, such as the potato tuber, where a largely parenchymatous organ of substantial size is produced by enlargement of the primary structure without secondary thickening. Usually, however, massive parenchymatous storage organs which can be used as food are provided only by those plants which show an anomalous form of secondary thickening in which the cambium produces mainly parenchyma and only a few specialized conducting cells. It is plants of this type, uncommon among wild plants, and belonging to only a very few species, which have been selected and developed in cultivation to produce root crops for animal feeding and stem and root vegetables for human consumption.

### STRUCTURE AND FUNCTION OF FLOWER AND FRUIT

The flower can be regarded as a specially modified shoot. Biologically it is essentially a device which first allows sexual reproduction to take place with as great a degree of certainty as possible, and with the greatest economy of material; it then provides, in the developing fruit, conditions in which the zygote can grow, at the expense of material provided by the seed parent, into a multicellular embryo with often substantial food reserves; finally, as a mature fruit, it provides means by which the embryo, by now dormant, can be dispersed, in the seed, so that this may be able to germinate and the embryo establish itself as a new plant in an area where conditions for growth are favourable. These different functions may be served in a very great variety of different ways, and flowers and the fruits which develop from them show an extraordinary multiplicity of form and structure. The floral parts may be many or few, free or joined, borne spirally or in whorls; the ovules, and consequently the seeds into which they develop, although always produced in some form of closed ovary, may be few or many, borne on the walls of the ovary or centrally; the ovaries may be numerous, each formed from one carpel, or single, formed either from one carpel or from the fusion of two or more carpels, with a single style or with separate styles, above the base of the other floral parts or sunk into the receptacle. Many of these variations in flower structure are adaptations to ensure pollination. Flowers pollinated by wind-borne pollen need to produce large quantities of pollen and to have large exposed stigmatic surfaces; usually self-pollination is avoided by the anthers and stigmas maturing at different times, but there may also be some form of partial or complete *self-incompatibility,* where pollen tubes grow slowly, or not at all, in styles of the same genotype. Apart from these requirements, wind pollination makes few special demands on flower structure. Insect pollination is usually more economical, in that pollen is conveyed directly to the stigma, but it involves special adaptation of the flower. This must be brightly coloured, and rendered conspicuous either by large size or by the close massing of small flowers; attractive by scent or by providing nectar; and constructed in such a way that the part of the insect which has come in contact with pollen in one flower will touch and deposit pollen on the stigma of the next. The adaptation is usually to one particular class of insect; thus flowers pollinated by flies are usually shallow and open in form, while those pollinated by bees are often deep tubular flowers. Normally the same devices to avoid self-pollination are found as in wind pollinated types, but there are some plants in which self-pollination is usual.

The fruit develops from the ovary, and its structure is therefore determined by that of the ovary; beyond this it may show a wide variety of form and behaviour. A single-seeded fruit is normally indehiscent and acts as the *diaspore,* that is the structure which is dispersed naturally, or, in the case of a cultivated plant, the structure which is intentionally sown. The fruit wall may be thick or thin, leathery or woody, smooth or provided with spines or hooks to aid external carriage by animals, winged or plumed to aid wind dispersal, or provided with some form of float to aid dispersal by water. If many-seeded it may break into one-seeded pieces, or dehisce to set free the individual seeds, splitting at the apex or down the sides, sometimes explosively, or open by pores or teeth so that seeds are shaken out by wind-induced movement of the parent plant. Or the fruit wall may be partly or wholly succulent, consisting of edible parenchyma usually with a high sugar content, so that it is attractive to animals, which in eating it disperse the seed which is either discarded or passes undamaged through the alimentary canal.

### CLASSIFICATION

The wide variations in the structure of inflorescence, flower, fruit and seed are the main features that are made use of in classifying the flowering plants, but there are equally wide variations in the vegetative structures, in growth habit, in leaf type and arrangement, in adaptation to different conditions, in the biochemical reactions which take place and in the substances which they produce.

The first rough-and-ready classification of plants was on growth habit, dividing them into trees, shrubs and herbs, but as more and more plants were described and studied in more detail (i.e. as botany developed as a science), the need for a more complete and exact system of classification became evident. The first readily workable scheme was the 'sexual system' proposed by the Swedish botanist Carl von Linné in 1753: at that time Latin was the accepted language for all science, and his name is usually latinized as Linnaeus. His system was based on flower structure only, and depended almost entirely on the number of stamens and the number of carpels; it was thus a purely artificial classification. Linnaeus also introduced a new system of naming plants. This was the *binomial system,* in which each plant was given a name consisting of two Latin words only, instead of the long descriptive names which had previously been used. Linnaeus' method of naming plants still remains in use, and Latin remains the language of botanical classification, but the artificial 'sexual system' has been replaced by a more natural system of classification which, by taking into account as many characters as possible, attempts to group together those plants which are in fact related in that they have evolved from common ancestors. The appreciation that present-day plants are the result of a long and still continuing process of evolution has given modern *taxonomy,* the study of plant classification, different aims from those of the earlier classifications which could regard plant types as the immutable products of 'Special Creation'.

A satisfactory classification must try to provide a scheme which not only represents the real relationships of plants, but which is also readily usable for identifying and indexing different plants, with the individual named categories (*taxa*) clearly enough defined, widely enough accepted, and permanent enough for the classification to be usable by botanists and by all those concerned with plant science as agriculturalists, foresters and horticulturalists throughout the world. No classification can satisfy completely all these different criteria, but the relationships of most plants are sufficiently clear for a consensus of botanical opinion to be agreed on a workable system. Classification and nomenclature are now governed by regulations drawn up by a series of international botanical congresses, and although, unfortunately, it is still occasionally possible for different names and different groupings to be used in cases where there are different opinions and different interpretations of fact, the system on the whole works well.

The system uses a hierarchy of categories, of which the main ones are class, order, family, tribe, genus and species. Thus the flowering plants as a whole are treated as a *class,* the *Angiospermae* (enclosed-seeded, from the seeds being produced within a closed ovary). This is clearly separated into two *sub-classes,* the *Monocotyledonae* and the *Dicotyledonae,* which differ in a whole series of characters beside that of cotyledon number. Thus, for instance, monocotyledons have flowers typically with parts in threes, with perianth not separable into calyx and corolla, stems with scattered vascular bundles and no secondary thickening, roots polyarch and leaves usually parallel veined; dicotyledons, on the other hand, have flower parts frequently in fours or fives, distinguishable calyx and corolla often present, stems with bundles in a ring, with typical secondary thickening developing, roots usually with few protoxylem groups, and leaves net veined.

The dicotyledons, the larger of the two sub-classes, can be divided into some forty *orders,* among which the *Rhoeadales* may be cited as an example. Members of this order, which takes its name from a classical name of a poppy, are mainly herbaceous plants with alternate exstipulate leaves, flowers in racemes with parts often in twos or multiples of two, calyx and corolla distinct and with separate segments, ovary superior and formed from two or more joined carpels, with parietal placentation and seed-reserves usually oily. Orders are large groups, with necessarily rather a wide range of characteristics; they are thus a category of more interest from the point of view of phylogeny than of practical classification, and it is mainly the next smaller category, the *family*, which is used for this purpose. Thus the order *Rhoeadales* contains six families, of which the two largest are the *Cruciferae* and the *Papaveraceae*. The *Cruciferae* have flowers with two pairs of sepals, four petals, six stamens, the ovary formed from two carpels and divided into two chambers by an outgrowth of the placentas, no latex and seed non-endospermic, whereas the *Papaveraceae* have two sepals and four petals, stamens usually numerous, ovary formed from two to many carpels, unilocular; latex is present in special latex vessels extending throughout the plant and the seed is endospermic. These two families illustrate the two possible types of family names; both names are adjectives, qualifying the understood feminine plural noun *plantae* (plants) and therefore taking the appropriate Latin adjectival feminine plural ending; *Cruciferae* means 'cross-bearing plants', from the cross-like arrangement of the petals, i.e. it is named from a character of the family, while the *Papaveraceae* is named from one of the important genera in the family, and means 'Papaver-like plants', *Papaver* being the name of the poppy genus.

Families are sometimes divisible into *sub-families,* and usually into *tribes,* but except where the family is large, or a detailed study is being made, the next category of importance is the *genus* (plural genera). Genera may be small or large, comprising from one species in a monotypic genus to several hundred species. The *species* (spelt the same in singular and plural, but conveniently abbreviated to sp. and spp. respectively) is the basic unit of classification; it is difficult to define exactly and comprehensively but usually all plants which can intercross to produce fertile offspring can be regarded as belonging to one species. All species belonging to the same genus have very many characters in common, and sometimes they will intercross, but usually to produce offspring which are sterile. The binomial system of nomenclature, introduced by Linnaeus, uses the name of the genus plus a second name to provide the proper name of the species. This second name is referred to as the trivial name or specific epithet. Thus in the family *Cruciferae* there are some 230 genera including, for example, *Brassica,* with sepals erect and fruit valves one-nerved, and *Sinapis,* with sepals spreading and valves three-nerved. *Brassica* has some eighty species including the cabbages and turnips, *Sinapis* five species including white mustard and charlock; *Brassica* and *Sinapis* species will not normally intercross. Among the *Brassica* species are *Brassica oleracea,* the cabbage species, having outer stamens with straight filaments, the flowers large and the unopened buds overtopping the open flowers in the raceme, and *B. rapa* (the generic name may be abbreviated to its initial letter when it has already been given once in full), the turnip species, with outer stamens having short curved filaments and flowers small with the buds below the level of the uppermost open flowers. *B. oleracea* and *B. rapa* will intercross, but only rarely and with difficulty, and the progeny are sterile unless chromosome doubling takes place. On the other hand different individual plants belonging to one species usually intercross readily to give fertile progeny; thus all members of the cabbage species will intercross even if they are of very different agricultural type, and the offspring will be fertile.

Latin proper names of species are usually printed in italics, and the botanical convention is that only the generic name has a capital. The generic name is a noun; sometimes it is the classical name of the plants, in which case it retains its original gender; more often it is a new word, adapted or made up in Latin form, and in this case its gender is decided by the botanist who first publishes it. The specific epithet is usually an adjective, again either an original Latin adjective or an invented one, and must then agree with the noun and show the correct adjectival ending according to the gender of the generic name; occasionally it is a noun in apposition, in which case it is invariable.

Difficulties can arise in the naming of species; for example a newly discovered plant may be given different names by different botanists. A full citation therefore includes the name of the authority, usually abbreviated; the cabbage species, named by Linnaeus, is in full *Brassica oleracea* L. Only one name is valid; this is the one that was published first, the starting point being Linnaeus' *Species Plantarum* published in 1753, and the international code lays down exactly what constitutes valid publication; other specific names are synonyms. Opinions may differ on the extent of a genus and a botanist may consider that a species originally named in one genus is better placed in another one; in such a case the original specific epithet must be retained, and the original authority quoted in brackets. Thus common chickweed was named *Alsine media* by Linnaeus, but has now been transferred to *Stellaria* as *Stellaria media* (L.) Vill. There may be other reasons why names need to be changed, as for instance where an earlier publication has been overlooked, and a later and therefore

invalid name has come into use. This is inconvenient, but it is generally considered that the advantages of the rule that the oldest name
has priority outweigh its occasional disadvantages. More serious
difficulties arise where the original description of the plant named
was obscure, so that there is doubt as to which species the name
properly refers to; this type of difficulty does not usually occur where
the type specimen has been preserved in a herbarium, but this is often
not the case, and it must be admitted that complete uniformity of
naming has not been achieved, and there are still some plants differ-
ently named by different authorities.

For wild plants identification of the species may be sufficient, but
even with these there may be a wide variation of types within the
species and more detailed classification may be required. *Subspecies*
(abbreviated as subsp.) is used for a taxon within the species which is
not merely separable on some morphological character, but is in
some way separated by geographical, ecological or genetic barriers
from other subspecies of the same species, so that although it can
cross with these, it normally remains distinct. Where this criterion is
not satisfied, the term *varietas* (abbreviated var.) may be used. This
term is sometimes anglicized as 'variety', but this usage involves the
danger of confusion with the quite different sense in which the word
variety is used in connection with cultivated plants. The subspecies
and the varietas are designated by Latin adjectives added to the
specific name, and the full citation includes the authority. Where a
species is separated into two or more subspecies, the type originally
described has the same sub-specific epithet as that which forms the
specific name, and does not need an additional authority; thus the
common white clover, extending over the greater part of Europe, is
strictly *Trifolium repens* L. subsp. *repens,* to distinguish it, where this
is necessary, from *T. repens* L. subsp. *prostratum* Nyman, a form with
small leaflets and densely hairy petioles, found only in certain parts of
southern Europe.

The scientific, binomial system of nomenclature may appear to
hold little interest for the practical farmer, yet it is important to the
student for various reasons. First, it assists towards an understanding
of the relationships of plants, which often have some bearing on their
agricultural characteristics. Secondly, the scientific names are of
value because they are fixed and definite, unlike the so-called common English names, which may vary from one part of the country to
another. For example, a very common pasture weed is known in
various parts of the country as ragwort, ragweed, staggerwort, stinking billy, stinking weed, yellow-weed, etc. As a result of this diversity,
a native of one area might not be understood in another area when

referring to the weed by the name common in his own locality. The botanical name of the weed, *Senecio jacobaea,* on the other hand, is not only used by botanists in all parts of the country, but is also used internationally. A further complication in the use of common names is the fact that one common name may be used for different species in different areas—for example, the name lakeweed is used in some districts for the species *Polygonum amphibium,* whilst in other districts it is applied to the species *Polygonum persicaria* (also known as redshank or persicaria).

It is of course only native plants, or ones that have been introduced for a long time, that have an English name. Many garden plants are exotics which are always referred to by their Latin names. A common name, if there is one, probably exists only in the language of the region from which the plant comes, and may well be unknown to anyone in this country. Moreover, there are cases where even common plants present difficulties when one wishes to name the species as a whole: in discussing names we have used 'the cabbage species' as the English equivalent of *Brassica oleracea,* but this could be ambiguous, since the species includes not only the plants we normally think of as cabbages, but also brussels sprouts, cauliflowers, some but not all of the plants known as kales, and various other cultivated forms, together with the wild sea cabbage (but not seakale), and excludes Chinese cabbage, Isle of Man cabbage, and a number of other and unrelated plants in which the word cabbage forms part of the English name.

# 2

# CROP PLANTS: GENERAL

ORIGINS

The growing of crops is essentially the controlled exploitation of photosynthesis. Animals have always depended on plants, and man, during the long stages of his development up to the close of the Palaeolithic period, was dependent upon wild plants collected and upon wild animals hunted for food. It was only in a few favourable areas that the wild flora and fauna was such as to provide man with sufficient readily available food throughout the year, and man was therefore a relatively rare animal. Only with the beginnings of agriculture, the bringing of suitable plants into cultivation, and the more or less contemporary domestication of suitable animals, did the position change. Agriculture probably developed independently in a number of different areas at different times, but the evidence suggests that the agriculture of Europe and Western Asia, which has by now profoundly influenced that of the whole world, originated near the eastern end of the Mediterranean some 10 000 years ago.

In some areas, such as parts of the Jordan valley, there existed natural stands of wild wheat and barley. These are annual mono-cotyledons adapted by their large self-burying diaspores to conditions where a long dry summer prevented the growth of any perennial vegetation other than widely spaced trees, and where a fair winter rainfall allowed good growth of those annuals, which by virtue of their large seed reserves were the strongest competitors in the autumn race for establishment. These large seed reserves, capable, to judge from present stands in the area, of providing some 500 kg/ha of edible dry matter of high digestibility, must have been quite outstanding sources of human food. They were expoited by local populations and eventually, with the discovery that the seed could be sown and similar plants raised in areas where they did not occur naturally, became cultivated crop plants. These cultivated plants were spread by migration of peoples and interchange with adjacent cultures; other plants were brought into cultivation, including first the large-seeded legumes, and later leaf, stem and root vegetables and succulent fruits, so that eventually practically all man's requirements for vegetable products were met from crops rather than wild plants. The food requirements of the domesticated animals continued for a long period to be met by wild plants, supplemented by straw and other residues from crops, and it was only later, with a highly advanced type of agriculture, that crops grown specifically for animal feeding became important.

In addition to those crops which provide food for man directly and through animal products, other plants were cultivated and developed to provide fibres, drugs and other useful materials; crops of this type are at present little grown in Britain and will therefore not be discussed in detail here. Timber requirements, which are largely obtained from what are essentially little-modified wild woody plants, may be excluded as belonging to forestry. It is thus primarily the food crops, whether conventionally classified as agricultural or horticultural, which concern us here.

### FOOD REQUIREMENTS OF MAN

Human beings on average need a daily intake of digestible organic matter with an energy content of some 11 MJ, that is to say the equivalent of some 700 g (dry weight) of digestible carbohydrate per day, or about 250 kg per annum. The type of carbohydrate is relatively unimportant, providing that it is digestible, and part of it can be replaced by fats or oils, which, with an energy content of some 38 kJ per gram as against 17 for carbohydrate, will enable the total dry weight requirement to be reduced. Carbohydrates and fats alone are of course insufficient; human food must also contain sufficient protein. The proportion varies, but as a standard figure it may be said that some 14% of the digestible dry matter should be protein; some apparently satisfactory adult diets contain substantially less than this, but the desirable figure for children may be up to 20%. The quality of the protein is also important. Proteins are very large molecules built up from amino acid groups; 26 different amino acids may occur in proteins and of these 8 (10 in children) can be described as essential in that the human metabolism is not able to produce them by conversion of other compounds. Moreover, amino acids are not stored as such in the human body; consequently the food eaten at any one time must contain the essential amino acids in the correct proportions. Other compounds are also essential in the human diet; these are the vitamins, which although required in much smaller quantities than the amino acids, are equally necessary. In addition, certain mineral elements must also be present.

These requirements for a satisfactory human diet can be met from plant sources, but not always easily. Plants are built up of many different tissues, but parenchyma is the only major tissue available as food. Typical vegetative parenchyma of leaves, young stems and similar structures normally has in the fresh state a water content of some 80–90%, and of the dry matter some 25% or more consists of cell walls not digestible by man. The dry organic matter which is digestible by man may thus only be some 8% of the fresh weight, and to obtain the required 700 g of it would necessitate the consumption of about 8 kg of fresh plant material per day. Vegetative parenchyma of this type can therefore not provide a satisfactory diet on its own, and it must be supplemented by more concentrated types of food. The parenchyma of special storage organs, which may contain up to about 20% of digestible dry organic matter, and that of seed reserves with perhaps 80% clearly provide much more satisfactory foods. The majority of the plants grown as crops for human food therefore fall into these two categories of 'root-crops' grown for their vegetative storage organs, and cereal, pulse and similar crops grown for their seed reserves. The low dry matter vegetative parenchyma crops for human consumption are mainly the leaf vegetables, eaten more for their vitamin and mineral content, and to provide variety in the diet, than for their energy content.

Man's food requirements in terms of energy can thus be met satisfactorily from crop sources, but his protein requirements are more difficult. The majority of the crops have protein contents of from 8% to 12%, and a purely vegetarian diet will usually be deficient in protein unless substantial use is made of pulse crops, which have protein contents of 25% or more. Moreover, vegetable proteins are frequently of low biological value, in that their amino acids are not present in the proportions required in human nutrition, and there is frequently a relative deficit of particular amino acids such as lysine. Normally therefore the human diet is based partly on animal products which have a higher content of protein of higher biological value.

## FOOD REQUIREMENTS OF ANIMALS

The use of animal products for human food introduces a further step in the biological food chain, and there is reduction in efficiency from the point of view of energy and quantity of organic matter. The amount of these obtainable from the animal will be, as a rough approximation, one-tenth of that in the plant material which it consumed. It is, of course, only where animals are being fed on plant material which could have been used directly for human food that this loss of efficiency is of any importance. When, originally, it was wild animals that were being hunted and their meat utilized, and later when domesticated animals were being fed largely on wild herbage and foliage, there was no competition with human needs, and all animal products were a net gain. Even in the more advanced type of agriculture of the present day the competition is only partial. The majority of domesticated animals providing meat and milk are ruminants. These are characterized by a digestive system which differs from that of human beings in three important ways: (1) the alimentary canal is much larger in proportion to body size, so that bulky low-dry-matter plant material can be utilized; (2) bacteria present in the rumen are capable of breaking down cellulose, and the products can be assimilated by the animal; and (3) the bacteria are able to synthesize amino acids which can be assimilated, so that providing the intake of combined nitrogen is sufficient, it need not all be in the form of protein, and the proportion of the different amino acids present in the food is unimportant.

Thus ruminant animals can make use of plants in which the main component is unconcentrated typical vegetative parenchyma which is not suitable for human food. They are not able to digest lignin, but the larger alimentary canal means that a higher proportion of fibre can be tolerated. Thus in an advanced type of agriculture, where crops are grown for feeding animals, these crops can be different from those grown for human food. They are normally grasses and herbage legumes, which may be higher yielding or cheaper to grow than crops for human consumption, and indeed can often be grown under conditions where no such crop would be practicable. Moreover, by-products such as straw, which are a necessary accompaniment of the production of human food crops, can be utilized in many cases. It must however be noticed that with intensive animal production, where maximum production per animal or per hectare is aimed at, such crops and by-products become inadequate, and more concentrated foods must be fed in addition. These are storage organs and seed-reserve foods, and there is thus some possible competition with the use of these for human food. The same considerations apply to the feeding of non-ruminant domesticated animals, particularly pigs and poultry, where again the plant products required are essentially similar to those used for direct human consumption.

TYPES OF FOOD CROP

The distinction between the types of crops grown for direct human consumption and those grown for animal feeding is thus by no means absolute. In general, almost any crop grown for human use can be fed to animals, if it is economic to do so; the only really distinct fodder crops are those with too low a percentage of digestible nutrients or too high a fibre content to be acceptable for human food. Even this presupposes that the crop products are to be eaten as grown; it is for example technically feasible to extract from herbage crops a leaf protein concentrate which would be a satisfactory ingredient of a human diet, the high-cellulose residue being still usable for the feeding of ruminants. This process is not yet used on a large scale, and there are difficulties with the palatability of the protein product, but should it come into general commercial use, it might be said that the distinction between the two types of crop had been broken down. From the point of view of the grower and user there is however still the distinction that crops for human consumption are commonly cash crops, valued not merely for their content of digestible nutrients, but for their acceptability to the consumer or processor. In such cases consumer preferences in terms of size, flavour, texture, appearance and many other characters will be the criteria. On the other hand, crops grown for animal feeding will either be retained on the farm or will be sold, eventually, to other farmers, and will be judged solely on their merits as producers of meat or milk. In their case the content of digestible nutrients is (providing that they are free from any toxic substances) the over-riding criterion.

Feeding value can be judged first from the dry matter content, and then from the digestibility of the dry matter. This will vary according to the type of animal to which the crop product is being fed, and can only be determined with complete accuracy by actual *in vivo* feeding trials, but for ruminants a satisfactory approximation can be obtained by much simpler *in vitro* determinations, using sheep rumen liquor. The figure may be expressed as digestible dry matter (DDM), but this will include the soluble mineral content of the dry matter, and the digestible organic dry matter (DODM) is usually preferred. The percentage DODM is often referred to as the D-value, and provides a useful figure for comparing the value of the dry matter of different crops, or of the same crop at different stages; it may be supplemented by figures for percentage fibre and percentage soluble carbohydrates, since these will affect palatability. The useful energy content of the food will depend on the proportions of the different substances in it, but for materials low in fats (i.e. most crop products other than oily seeds) the approximate metabolizable energy concentration, expressed as MJ/kg can be obtained by dividing the D-value by 6·3. Normally it will be necessary to know also the percentage of protein in the dry matter.

Food crop plants are normally grown for their parenchyma tissue, and this can be separated into three types according to its dry matter content. The three types tend to occur in different parts of the plant, and food crops may therefore be classified into groups according to the part of the plant which is utilized.

1. *Leaves and young stems.* These consist largely of typical low dry matter vegetative parenchyma; older stems usually have too high a proportion of lignified tissue to be useful for food. Crops grown for their leaves and young stems provide mainly food for ruminant animals, since the proportion of water and of cellulose is usually too high for them to form a staple human diet; some plants of this type are however used as vegetables. Such 'greens' or leaf-vegetables are usually eaten in comparatively small amounts—mainly for their vitamin C and mineral content rather than as a source of energy—and are horticultural rather than agricultural crops. As crops used for animal feeding, the plants grown for their leaves and young stems provide non-concentrated foods; the water-content is usually high, and the food is available only for immediate consumption unless it is specially preserved by drying, or as silage.

The method of utilization of crops of this type depends largely on the growth-habit of the plant; plants with upright stems and aerial buds can usually only be cut or grazed off once (e.g. kales), while those with short or creeping stems and buds at or near ground-level will withstand repeated cutting or grazing (e.g. grasses and clovers). In both cases stock can be fed directly (i.e. the plants grazed), so that costs of immediate utilization are reduced to a minimum. The period over which such grazing can be carried out is dependent in Britain mainly on the ability of the plants to withstand frost and the treading of animals in wet soil conditions; in hot, dry climates it will be limited by drought conditions, although during drought periods animals may continue to feed on the dead, dry herbage.

Crops of this type, where leaves and young stems are eaten, thus normally form the main food of herbivorous animals, and it is only to provide for special conditions that the more costly crops of types 2 and 3 below are utilized. Such conditions arise when the leaf and young stem crops are not available and also when the animal's maximum intake of these bulky crops, either fresh or preserved as hay or silage, does not contain sufficient digestible dry matter to provide the required live weight gain or milk yield.

2. *Special storage organs*. The vegetable storage organs found in food crops are mainly stems and roots in which there is a massive development of parenchyma in which food material is stored. They thus provide moderately concentrated foods, usually storable for some months and available for winter consumption. They are produced mostly by a few very specialized, long-cultivated plants, since species which form large volumes of edible parenchyma are rare. The majority are biennials, harvested at the end of the first season's growth, when the storage organ has reached its full size and while the plant is still in the vegetative stage. If conditions are such that flowering occurs in the first season, the value of the storage organ is much reduced. In potatoes, and the structurally similar Jerusalem artichoke, the plant is a perennial, dying down each year, and perennating only by means of underground stem-tubers; in these, production of the storage organ is largely independent of flowering.

Most of the crop plants of this type (2) have a storage organ which is mainly swollen tap-root, but the hypocotyl and the lower part of the true stem (epicotyl) may contribute to it. In kohlrabi the swollen part is, however, wholly stem, the root and hypocotyl being slender and woody; in potatoes the storage organ is a true tuber—that is, the enlarged apex of an underground axillary stem. Anatomically the storage organ of all these crops consists essentially of a mass of parenchyma, but the way in which this originates shows great variation. It may be primary tissue, as in potatoes; more commonly it is secondary tissue, derived either from a single cambium, as in turnips and carrots, or from a series of concentric cambia, as in mangels.

Crop plants of this type require to be grown at wide spacing to allow full development of the storage organ, and they are therefore usually expensive crops to grow. If grown for animal consumption, only those with storage organs mainly above ground can be fed off *in situ*, and the remainder (including mangels, which, although mainly above ground, are chemically unfit for feeding before winter, and are not frost-hardy) must bear the cost of lifting, storing and feeding in addition. This cost of handling is necessarily high per unit of food owing to the comparatively high proportion of water present (dry matter ranges from about 8% in turnips to about 24% in potatoes). Originally garden crops for human consumption, plants of this type came into field use at a time when labour costs were comparatively low, and the need for winter feed for stock pressing. The introduction of turnips and swedes as winter feed for stock may indeed be said to have effected a revolution in British agriculture. With increased labour costs, however, and with the development of alternative winter foods, such as kale and grass silage (i.e. foods derived from the less expensive leaf and young stem crops of type 1), such root crops have become relatively less useful as animal foods. The development of efficient selective weed-killers has also reduced the necessity for their inclusion in the rotation as cleaning crops. There is therefore a tendency to reduce the acreage of these crops grown for stock, and where they are grown to utilize mainly those, such as turnips, which can be fed off *in situ*, or alternatively, where storage is required, those of which the harvesting can be mechanized. Potatoes and sugar-beet are normally grown only as direct cash crops for human consumption (in the latter it is the sugar only which is so utilized, the cellulose, etc., remaining as a by-product for stock food); carrots and parsnips, formerly grown to some extent for stock food, are now almost exclusively grown as vegetables for human consumption, and even swedes owe some of their popularity to the fact that they may be saleable for this purpose.

Storage of all these crops depends on the winter-dormancy of the plant, and cannot be continued after growth recommences in spring.

3. *Seed reserves.* Seed reserves, present either in endosperm or cotyledons, normally have a very low percentage of water, and give highly concentrated foods which can be readily stored. The low cellulose content of many seeds makes them suitable as human food, and the fact that they remain dormant indefinitely if kept dry means that they can be stored for long periods.

Amongst the various large-seeded plants grown for their seed reserves the cereals are pre-eminent; these are grasses in which the bulk of the large seed consists of endosperm, parenchyma tissue densely filled with starch grains. Being upright annuals, with large seeds giving ready establishment, they can be comparatively cheaply grown in close stand, without interplant cultivation. The fruits, ripening all at the same time, and borne well above the ground, are readily and cheaply harvested.

Seed reserves are important sources of proteins and fats as well as of starch, but in this and other temperate climate countries with a comparatively high standard of living the tendency has been for proteins and fats for human consumption to be derived largely from animal sources. Where vegetable fats are required, oil seeds are imported from warmer climates, and the residues after oil extraction provide concentrated protein foods for stock. Oil and protein seeds are therefore of somewhat lesser importance amongst British crops, but changes in world conditions have resulted in the imported products becoming less readily available and relatively more expensive and there is therefore increasing interest in the possibilities of home-grown protein and oil crops.

4. *Succulent fruits*. These represent a special case of the development of edible parenchyma. Typically part or the whole of the pericarp, or of the pericarp and receptacle, becomes succulent. The structure shows a wide range of variation depending on the original structure of the flower and the extent to which succulence develops. Succulent fruits include a number of forms such as marrows, cucumbers and tomatoes commonly regarded as vegetables, as well as the wide range of types classified as fruit from the dietary rather than the botanical point of view. These latter have mostly a high water content; the main nutrients are usually sugars and organic acids, and most are eaten for their refreshing flavour and for their vitamin C content. Some exceptions occur, thus for example among subtropical fruits bananas have a fairly high starch content, in olives oil is present in the pericarp, and in avocados oil and protein.

The majority of succulent fruits provide only a relatively low yield of edible material, and thus are expensive crops for human consumption only. Many are long-lived perennial crops, slow in coming in to bearing, and expensive to cultivate; since the produce is usually very readily damaged, mechanical harvesting is only rarely possible. The storage and marketing of fresh fruits often presents difficulties; ripening is an irreversible process involving conversion of starch and organic acids to sugars and a progressive softening of cell walls. Usually a certain stage of ripeness must be reached before harvest, and accurate control of temperature and of the storage atmosphere may be necessary for the fruit to reach the consumer in good condition, particularly where storage is prolonged.

### CROPS GROWN FOR SPECIAL PRODUCTS

Some crops are grown for products which form part of human diets, but which contain little of direct nutritive value. These are the herbs, spices and beverage plants, grown because they produce chemical compounds of attractive flavour or pleasant physiological effect. These compounds are the essential products, and the type of tissue and the part of the plant in which they occur is quite unimportant; in different crops of this type root, rhizome, bark, leaf, flower bud, stigma, fruit, seed and aril are used. All are consumed in relatively very small amounts, and non-parenchymatous products are either employed in powdered form, or essences or decoctions are used. The chemical compounds produced are often of relatively simple structure; terpenes, alcohols and esters occur commonly. In some cases the main constituent can be cheaply synthesized, but the natural product, containing a range of associated compounds, is usually preferred as providing a more attractive flavour. In a number of examples the flavouring compounds are not developed in the growing plant, but are produced as a result of some form of post-harvest fermentation.

The main spices and beverage plants are tropical or sub-tropical crops; the herbs and temperate flavouring plants are mainly small-scale specialist crops, and occupy only a small area in Britain. No hard-and-fast line can be drawn between such crops grown for the flavouring of foods and drinks and those grown for similar special compounds used in different ways: medicinal herbs and drug plants, tobacco. Rather more distinct are crops grown for special products used for technical purposes, such as plants producing rubber, dye-stuffs, perfumes, waxes and varnishes. To complete the list of crop types there are those grown for their structural components, fibre- and timber-producing crops, and those used for ornamental and amenity purposes.

THE YIELD OF CROPS

The plants employed as crops will normally be those that give the highest yield per hectare of usable product. Yield can be defined in a number of different ways. *Biological yield* is the total gross amount of material produced by the crop, above and below ground, including plant parts that die and are lost before harvest; it is thus a figure which is not in itself of direct economic importance, and one which is difficult to assess accurately. *Economic yield* is the yield of that particular part of the plant for which the crop is being grown; in a forage crop it may be the total yield harvestable by whatever means of harvesting is adopted, but in a crop grown for fruit or seed it may be a relatively small fraction of the harvestable yield. Partition of the products of photosynthesis between the different parts of the plant is thus an important criterion in determining the suitability of a plant for crop use. A desirable crop will, in general, be one in which the economic yield is a large fraction of the biological yield. A crop in which partition is such that the economic yield forms only a small fraction of the biological yield will only be worth growing if the product is of very high economic value.

These different types of yield can be expressed in various ways. A cash crop yield will usually be expressed as weight of saleable product in the form in which it is sold. Yield of a crop grown for animal feeding

may be expressed as fresh weight or as dry weight, or more signifi-
cantly as weight of digestible dry matter, or of digestible dry organic
matter. In all cases, quality will be important; thus in forage crops the
yield of protein may be a useful figure; in cash crops for human
consumption the yield in particular quality grades, and in crops for
extractive processing the yield of a particular end product, as for
instance the sugar yield from sugar beet, may be a more useful figure
than total yield.

*Factors affecting yield.* Biological yield, and hence the other
measures of yield which derive from it, is determined by very many
inter-related factors of the plant and the environment in which it is
grown. The primary factor is the amount of radiant energy available,
and the efficiency with which it is used in photosynthesis. Radiant
energy can only be utilized if it is intercepted by leaf or other green
tissue; a high-yielding crop must therefore be one which can be grown
in such a way as to provide complete ground cover over as large a part
of the growing season as is possible. This in turn will necessitate the
availability of sufficient water to compensate for the unavoidable loss
by transpiration from this complete leaf cover. Further, since in most
temperate crops leaves become light-saturated at relatively low light
intensities, the leaf cover must be arranged in such a way that the light
is shared amongst a number of leaves. Crops with a high *leaf area
index* (LAI: ratio of area of foliage to area of ground) make more
efficient use of the available light energy than those with a low LAI.

For a high rate of photosynthesis to be maintained, there must be
an adequate *'sink'* to which the elaborated material can be translocated. This may consist of rapidly growing parts of the plant, which
utilize the materials immediately, or of a storage region, either a
vegetative storage organ or developing seeds. Herbage crops for
animal feeding are normally managed in such a way that mature seed
is not produced; there is little storage, and the products of photosynthesis are used immediately in new growth. The highest yielding
herbage crops will be those in which a high proportion of this new
growth is itself photosynthetic, so that leaf area increases logarithmi-
cally in the early stages. This type of growth cannot of course be
continued indefinitely: an optimum LAI is reached, and beyond this
point the lower leaves are partially shaded, and are therefore carrying
out less photosynthesis. Ultimately a ceiling LAI (say about LAI 4 or
5 for many broad-leaved crops such as clovers, about 11 or 12 for
common grasses) is reached, where production of new leaf area is
balanced by the dying off of lower leaves which are so heavily shaded
that they are below the light compensation point, and are losing more

material in respiration than they gain by photosynthesis. Herbage crops are of course normally harvested, either by cutting or grazing, before they reach this point.

Root crops, and similar crops with a vegetative storage organ, are usually biennials in which the greater part of the photosynthetic production which is in excess of the plant's immediate needs is stored during the first year. The crop is normally harvested at this point; if it is left the stored material is mobilized in the second year and used, together with the products of current photosynthesis, in the production of the flowering stem and ultimately of the seeds.

Crops grown for their seeds are usually annuals in which there is relatively little vegetative storage; photosynthesis in the earlier stages is important in providing a sufficient area of leaf and an adequate inflorescence, but the seed reserves, for which the crop is grown, are often largely derived from post-flowering photosynthesis. The essential requirements for high yield in these crops are thus an adequate number of developing ovules and a large and long-persistant leaf area.

The plant factors contributing to high crop yield are thus essentially a large source, i.e. a large leaf area, and a large sink, the two being accurately matched both in capacity and in timing. The sink must, for food crops, be a large volume of edible storage parenchyma with relatively little associated lignified tissue, or a series of active meristems giving rise to edible parenchyma in the form of leaves and young stems. These requirements are met by only a comparatively small number of plants, and it is these which have been brought into cultivation and developed as crops. This development, from the original wild ancestors, was in all its earlier stages a slow, unsystematical trial-and-error process, although over the course of centuries often very effective. Now, with the possibility of conscious organized selection by the plant breeder for the plant factors which contribute to high economic yield, it can be a much more rapid and even more effective process.

The main environmental factors controlling crop yield are the amount of radiant energy, discussed above, and the carbon dioxide content of the air, the temperature, and the soil conditions, including water supply and the availability of mineral nutrients. Only some of these factors can be controlled by the grower. Radiant energy in any particular area is of course not controllable in any economic way, although for special experimental and plant-breeding purposes plants can be grown under artificial light in controlled-conditions growth chambers. Tropical regions with a high incident radiant energy have a higher potential yield level than temperate regions. In any given area, the maximum yield can only be obtained by making the best possible use of the incident energy available, that is by growing a crop in such a way as to provide the highest possible effective leaf area index over as large a part of the growing season as possible. The carbon dioxide content of the air is economically controllable under glass-house conditions, where substantial increases in yield are obtainable with carbon dioxide enrichment of the atmosphere, but cannot be effectively altered for outdoor crops. Temperature again is controllable only under glass-house or protected-cultivation conditions, and for the unprotected open-air crop the most that can be done is to manipulate the timing of the crop so that growth takes place during the most favourable part of the year. Water supply can to some extent be altered by drainage, or, where an economic source of water is available, by irrigation, but the factor most readily changed is the supply of available soil nutrients.

While it is normally possible to grow a crop of some sort under almost any set of environmental conditions other than true desert or tundra, the general tendency of advancing agriculture has been to concentrate mainly on food crops for relatively good conditions. Such a crop is one which has a high intrinsic rate of photosynthesis and a long growing season during the most favourable part of the year. The partition of the products of photosynthesis will be such that a high proportion of the yield consists of utilizable edible parenchyma, and the crop will respond well to heavy manuring. This ideal type of plant is in general the object aimed at in the choice of crop plants, although clearly all the requirements cannot always be met. Thus for some purposes a crop is required which comes to maturity early; here the growing period is shorter and the yield is thus reduced. Such a crop will only be desirable where the increased value of the early produce compensates for reduced yield, or where early maturity either allows the plant to be cultivated in an area where climatic conditions do not permit the growth of longer-growing, higher-yielding forms (e.g. spring wheats in northern Canada), or permits of a second crop from the same land being taken during the one year. Again, the requirement that the period of growth shall be during the most favourable part of the year cannot always be met; it may be necessary to grow a crop at a less favourable period in order to provide fresh food for stock at a particular time, or to use it as a catch-crop. Sometimes the physiological behaviour of a plant is such that the period during which it can be grown is limited. Thus, for example, sugar-beet and related crops, which are biennials grown for the food reserves in the storage organ, are normally sensitive to low temperature during the seedling and early-growth stages. Sowing early enough to give maximum growth during the most favourable part of the year is therefore impossible, as such plants respond to the low temperatures to which the seedling is subjected by bolting—that is, by behaving as annuals, so that little or no food reserve is built up.

The nutrient status of the soil is usually the factor most readily controlled, and with increased use of fertilizers there is a tendency to use only crop plants adapted to conditions of high fertility. The great majority of newly-introduced crop varieties are ones which give high yields in response to heavy manuring, and crops and crop varieties which are suited only to conditions of low fertility are tending to become of less and less importance in British agriculture. This general tendency towards the use of high-fertility plants extends not only to arable crops, but also to grassland; even on poor hill grazings it will usually be considered better to build up the fertility to a level at which the higher-yielding grasses and clovers can be grown, rather than to try to find and grow the species which would give the maximum yield at the existing level of fertility. Such a tendency is of course dependent on the value of the crop increase being greater than the cost of the fertilizers. Readily available reserves of the raw materials are moreover not unlimited except in the case of nitrogenous fertilizers manufactured from atmospheric nitrogen, and here a very large input of energy is required.

*Ease of cultivation.* The cost of growing a crop is largely controlled by the amount of labour which has to be expended on it; plants with a high labour-requirement will only be grown if they yield a high-priced product, usually for human consumption, and the growing of such crops (e.g. hops, celery, strawberries) tends to be a specialized horticultural operation rather than a matter for the general farmer. Where crops are grown for feeding to stock or to provide low-priced human food, it is necessary that they should be grown cheaply. This means that as far as possible the plants must be readily established, that the cultivations required shall be simple and capable of being mechanized, and that the crop can be either fed off *in situ* if for stock, or harvested quickly and cheaply.

It might be thought that a perennial crop, in which the problem of establishment does not arise every year, would be an advantage; in fact, any advantages due to this fact are more than offset by the difficulties of maintaining high fertility and of keeping down weeds in perennial crops. Inter-row cultivations of standing crops are always more expensive and usually less effective than the thorough field cultivations which can take place during the interval between two annual crops; if hand-hoeing is involved, the expense may be prohibitive. As a result, long-lived spaced crops are found in this country only among specialized horticultural crops, such as hops and soft fruit. With close crops, where the plants are closely spaced with little or no bare ground between, the weed problem may not appear so serious, but it is still present. The problem of maintaining fertility is even greater, and it is only grass and grassland plants which are grown as such a permanent crop. Even here the tendency for fertility and yield to deteriorate, and for weeds to come in, is very marked unless the crop is being grown under particularly favourable conditions. In many cases, therefore, one-year or short-term leys are preferred to permanent grass, the extra cost of re-establishing the crop at frequent intervals being more than offset by the increased yield obtained.

With the partial exception of grassland, then, perennial plants play little part in British agriculture, and the great majority of crops are annuals, or biennials and perennials grown as annuals. Such crops are usually grown from seed; crops which must be propagated vegetatively are normally more expensive to grow, and the potato is the only common agricultural crop in which this method is used.

It follows that for a plant to be satisfactory as an agricultural crop it is desirable that it should produce seed readily and that it can be easily established from this seed. The part sown need not be a true seed in the botanical sense, it may be an indehiscent fruit or an even more complicated structure. It is therefore often convenient to use the non-committal term *diaspore,* which refers to the part of the plant naturally distributed or intentionally sown, whatever its botanical nature may be; diaspore thus covers the range of senses in which the word seed is commonly used in agricultural parlance. In crops which are grown for their seed reserves, that is ones in which the diaspore, whether it is true seed, indehiscent fruit or (as in barley for instance) fruit plus other parts of the inflorescence, is the normal product, there is usually no difficulty. In the case of crops normally grown for their vegetative structures, as for example in herbage and root crops, special seed crops must be grown, and the treatment of the crop when grown for this purpose may need to be quite different from that normally employed.

Many wild plants have very effective and rapidly acting mechanisms for the dispersal of their diaspores. This is of course a disadvantage in a cultivated plant and in most crops grown for their seed or fruit this mechanism has been much modified. Thus for example in linseed and in oil-poppy, where the diaspore is the true seed, the wild types have dehiscent fruits, but the cultivated forms have been selected for absence of this dehiscence mechanism, and the fruits remain whole until the crop is harvested and the seeds released by threshing. Similarly, the rapid disarticulation of the mature inflorescences characteristic of wild cereals, which resulted in complicated self-burying diaspores, has disappeared in the cultivated forms. These have simpler diaspores which remain attached to the much tougher inflorescence axis until well after full ripeness is reached, and are only separated on threshing. On the other hand some of the crops usually grown for their vegetative yields have not been so closely selected for convenient seed production; some of the cultivated annual lupins, for example, have fruits which dehisce very readily, with consequent pre-harvest loss of seed. In other crops not grown for their seed reserves the diaspore may be of inconvenient shape or structure, as in carrots, beet and some of the herbage grasses.

Such diaspores can sometimes be improved by mechanical treatment; thus in carrots, where the natural diaspore is a mericarp with one surface covered by long spines, it has long been the practice to 'mill' the threshed diaspores so that the spines are largely removed, and the carrot 'seed' as purchased by the grower is relatively free-flowing and can be drilled. In beet the natural diaspore is a complicated structure consisting of the fused remains of two or more flowers, each containing a single seed. The diaspore is thus not only of inconvenient shape, but may give rise to several very closely spaced seedlings and therefore necessitate expensive singling of the resultant crop. This difficulty can be overcome in beet either by mechanical treatmeant of the diaspore, or by the breeding of monogerm varieties, in which the flowers are borne singly and not in clusters. The new type of diaspore is satisfactory in that it is single seeded, but it is still often inconvenient in shape and irregular in size. Precision drilling and 'drilling to a stand' have become increasingly important in that they enable optimum spacing of plants to be achieved without expensive singling or transplanting, and such drilling necessitates 'seed' of uniform size and shape. Inconvenient diaspores are therefore often pelleted, that is, enclosed in an inert shell to give what is in effect an artificial diaspore suitable for precision drilling. Where the natural diaspore is of satisfactory shape, as for example the true seeds of kales, uniformity of size can be obtained by close grading of the seed, and pelleting is unnecessary.

Once the crop is drilled and has been established it will usually be of economic advantage if pre-harvest cultivations can be kept to a minimum. The former special value of root-crops grown in wide drills, which could be regarded as cleaning crops in that cultivations for weed control could be continued until a complete ground cover of leaf developed, has very largely disappeared. Control of weeds in the growing crop is now largely achieved by the use of selective herbicides, and resistance to suitable herbicides is therefore a desirable attribute of a crop plant.

Harvesting costs may form a large proportion of the total cost of a crop. Herbage crops for immediate consumption by animals may be grazed *in situ* and this will normally be the cheapest method of harvesting. Crops for storage or for human consumption will need to be cut, picked or dug; the process can normally only be fully mechan·-ized if complete unselective harvesting can be practised. Cereals and other crops grown for their seed reserves harvested at the fully ripe stage are normally combined, and for this plants are needed which are stiff-standing, without excessive persistent succulent foliage, with all plants and all seeds or fruits maturing at the same time, and with as long as possible a safe period between ripening and the beginning of natural shedding. Root-crops for mechanical lifting need to be of uniform size and shape, and with foliage that can be readily separated from the storage organ. It is with crops for human consumption that most difficulties arise. Thus many crops grown for use as vegetables necessitated selective harvesting by hand since they came to maturity over a relatively long period. Mechanized harvesting has been made possible by the development of special single-harvest varieties, as for instance in vining peas and in brussels sprouts. In these a sufficiently large proportion of the pods or sprouts reach the required stage at the same time for it to be possible to cut the whole crop, the required parts then being removed from the stems by machine. Such single-harvest varieties must show close uniformity of plant type, usually only possible with a self-pollinating plant or an F.1 hybrid, as well as a specialized pattern of growth.

This type of harvesting by cutting of shoots and subsequent machine stripping can be extended to perennial crops where above-ground parts are short-lived. Hops, with annual climbing stems, are harvested in this way; black currants, where fruit is borne on stems in their second year, can be so treated if alternate year cropping is accepted. Mechanical harvesting of fruit borne on older stems, as in most fruit trees, necessitates, where it is possible, some form of mobile shaking or picking device.

These developments in methods of cultivating and harvesting crops have necessarily influenced the crops themselves. Crops which in the past have been mainly small-scale market garden crops with high labour requirements have become field-scale farm crops; new varieties suited to the new methods have been developed. The selection of varieties has also been strongly influenced by changing methods of use; this again is particularly marked in crops grown for human consumption. Where previously almost all crops, other than those harvested as dry seed or fruit, were used fresh, or during the relatively short period for which they could be stored without deteriorating, a large proportion are now grown for processing. This usually involves freezing, canning or dehydrating: factory processes which demand uniformity of size, colour, consistency and biochemical activity in the produce, as well as predictable yields and times of maturity to allow efficient organization of the factory intake. Even where vegetables are grown for immediate consumption, without prior processing, they are often sold 'pre-packed' in such a way as to favour varieties which are of attractive appearance. All these changes in methods of marketing and utilization necessarily influence the characters required in the crop.

### THE CLASSIFICATION OF CULTIVATED PLANTS

The system of classification and nomenclature outlined in Chapter 1 has been primarily developed for the grouping and naming of wild plants. Its application to cultivated plants presents some special problems. There are no difficulties with the larger groups, and the few intergeneric hybrids are simply dealt with by combining the generic names, as for instance in *Rhaphanobrassica*. There may however be complications at the specific level; an interspecific hybrid, if it is known to be such, either because it is infertile, or because its hybrid origin was recorded, is named as such, with the generic name followed by ×, e.g. the common cultivated strawberry is *Fragaria* × *ananassa* Duch., a cross between two wild American species, *F. virginiana* Duch. and *F. chiloensis* Duch. In many cases however a plant now known to be a fertile hybrid was originally named as a good species, and the name is retained. Further, there are many crop plants which have probably developed as a result of complicated and unrecorded hybridization over a very long period of time, and which may contain genes derived from a whole range of original species. In such a case, if we were to apply rigidly the rule that all plants which can intercross to produce fertile offspring are to be regarded as conspecific, then all the contributing parents would have to be lumped together in one enormous species. Such a 'biological species' would be a very inconvenient unit for classification purposes and is therefore not generally used. There are also of course many cases in which the mode of origin of a particular cultivated plant from wild species has not been worked out. For all these reasons it is impossible to arrive at a wholly consistent and logical treatment of the cultivated species, and some differences of opinion and usage are inevitable.

These are however minor and perhaps temporary difficulties, and there is no question that the advantages of using standard specific nomenclature for cultivated plants far outweigh the disadvantages.

A species which is extensively cultivated may exist in many hundreds or even thousands of forms. These are the named varieties employed in agriculture and horticulture, and to them the technical term cultivar (abbreviated cv.) is applied in scientific literature, although for ordinary purposes the word variety is still commonly used. The cultivar is for most practical purposes the ultimate unit, which cannot normally be subdivided, but the range of type within a single cultivar will vary with the crop and the mode of propagation. In vegetatively propagated crops, such as potatoes or apples, the cultivar is a clone, and all the individual plants belonging to it have the same genotype, being derived vegetatively from a single original seedling. In crops grown from seed the range of type will depend on whether the crop is inbreeding or outbreeding. In a self-pollinated crop such as peas individual plants will be homozygous and essentially of the same genotype, so that the range of variation within the cultivar will be not much greater than that within a clone. In a cross-pollinated crop, and where the seed is produced as a result of open pollination, as for example with most cultivars of cabbages, individual plants will be heterozygous and of different genotypes, and may thus show a fairly wide range of type within the cultivar. If however the seed is produced by controlled crossing of two inbred (and therefore largely homozygous) lines to produce an F.1 hybrid, then all the plants within the cultivar will be heterozygous but will have essentially the same genotype and show a high degree of uniformity, as for example in many recent cultivars of brussels sprouts.

Cultivars of all these types are treated in the same way from the point of view of classification. Their naming is governed by the *International Code of Nomenclature of Cultivated Plants 1969*. The rules under the Code are mainly quite simple: the full name of a cultivar is the proper name, or the common name, of the species, plus a cultivar name. This latter should be a 'fancy' name, that is a real word in English (or the language of the country where the cultivar originates) or an invented word not in Latin form. The cultivar name may be enclosed in single inverted commas if this helps to make it more distinct; unlike the specific name it is printed in roman type, not italics. Thus one could use the forms 'Majestic' potato, or *Solanum tuberosum* Majestic; if necessary the subspecies or varietas can be added to the proper specific name, e.g. a particular cultivar of florence fennel is *Foeniculum vulgare* var. *azoricum* 'Perfektion'. The

Code makes various regulations and recommendations about suitable types of names, about translating them from one language into
another, about synonyms where it happens that one long established
cultivar is known under more than one name, and about what constitutes valid publication of a cultivar name.

Cultivar names are also governed in Britain by regulations made
under the Plant Varieties and Seeds Act 1964 as amended by the
European Communities Act 1972, which in fact delimit what can be
regarded as an acceptable cultivar. It should be noted that the Act
and regulations use the term variety throughout; this means exactly
the same as cultivar, and it is only in botanical works, where there is a
danger of confusing variety in this sense with the anglicized form of
varietas, that the use of the term cultivar is essential. For most crops, a
cultivar may only be sold if its name is included in the appropriate
National List (or with certain exceptions in the Common European
Catalogue). All cultivars in commercial use at the time when the list
was drawn up were included, but since then all new cultivars must
pass certain tests before being admitted to the list. These tests are for
distinctness, uniformity and stability; that is to say a new cultivar must
be clearly distinguishable in some way or other from all cultivars
already on the list, it must be uniform or vary only within an acceptable range according to the method of propagation, and must remain
true to type in successive years. Further, for some crops, a new
cultivar must have economic merit, that is its performance in terms of
yield or quality must give it some advantages. Provision is made for
the periodic revision of the list, and removal of cultivars which do not
satisfy these criteria. Cultivars which are so little used that no seeds
firm is prepared to bear the cost of maintaining them are automati-
cally deleted.

It will be apparent from this in fact much simplified account that
the status of a cultivar is very firmly fixed and regulated, and that
from the classification point of view the cultivar, once tested,
accepted and listed, presents no difficulties. We have seen also that
the species, although much less firmly regulated, is at least reasonably
clear in most crops. It is in the area between species and cultivar that
classification may present difficulties. The botanical concepts of subspecies and varietas are available, but are not really appropriate for
many of the cultivated crop plants. There are other Latin-named
categories which can be used; these are convarietas, which comes
above varietas, and subvarietas and forma, which come below, but
the same objections could be raised to these. If all these categories are
used the nomenclature becomes extremely complicated, and even
then the system is not really adequate to deal with a species like

*Brassica oleracea* where the number of agriculturally and horticulturally distinct groups of cultivars is very large. The position is still further complicated by the fact that the use of Latin-named botanical categories involves the acceptance of the principle of priority; the history of the various epithets used to name the groupings between species and cultivar is often obscure, the same epithet having been used by different authorities for different categories. Some of the groups which have been used are based on quite minor differences such as colour or puckering of leaves, others on differences in growth-habit and life-history which do correspond to different economic uses. However, economic uses tend to vary, and may now no longer agree with those of the time when the grouping was first proposed, and the name given. Moreover, the dividing line between groups tends to be broken down; frequently where two distinct groups of cultivars exist within a species they will be intercrossed by plant-breeders in an endeavour to combine the advantages of both, and the resultant new cultivars may not fall clearly into either of the original groups.

For all these reasons the position was taken up in earlier editions that Latin-named categories within cultivated species were better avoided, and that for most purposes the English-named groups of the type used in crop-husbandry provided a more satisfactory classification of cultivars. With the accession of Britain to the European Economic Community this position needs modification; clearly, an internationally acceptable classification is now necessary. For this there is no satisfactory alternative to Latin; the E.E.C. employs a classification using numerous Latin-named infraspecific categories and this has been accepted by Britain and is used in the *Plant Varieties and Seeds Gazette,* the official publication in which cultivars are recorded. It must therefore be adhered to, as a matter of practical necessity, where a detailed classification is required and where Community interests are involved. It is however essentially a compromise, adopted as a workable international system for official legal purposes, rather than an ideal biological classification.

This edition therefore gives, for reference purposes, the full Latin name for each of the crop plant groups, as adopted in the E.E.C. classification. In the general discussion, however, Latin names are usually confined to those for species; for types of crop within the species the simpler and more convenient English terms are used. It occasionally happens that the specific name adopted in the official E.E.C. scheme is not the one usually regarded as the most appropriate botanically; in such cases the E.E.C. usage is given for reference, but elsewhere the more usual name is employed.

### THE TAXONOMIC DISTRIBUTION OF CROP PLANTS

It is evident that a plant must meet a large number of very stringent criteria if it is to be a satisfactory crop plant. The total number of species of flowering plants existing in the world is well over a quarter of a million; of these only some 250 are cultivated on any important scale. Most of these have been in cultivation for many centuries; new cultivars of existing crop species are constantly being developed, but the bringing of a new species into cultivation is a rare event.

In Britain, and in temperate climates generally, the number of important crop species is even smaller. In Britain the figure (excluding forest trees and ornamentals) hardly reaches 100 species, of which some three-quarters are dicotyledons. The majority of these are relatively minor crops occupying only a small proportion of the total cropped area, and the important dicotyledonous crops belong to some twenty species only. As might be expected, many of these species are closely related, and all twenty derive from six families; the *Cruciferae, Chenopodiaceae, Umbelliferae* and *Solanaceae* providing the agricultural root and forage crops and the related root, stem and leaf vegetables, the *Leguminosae* the pulse crops and some herbage crops, and the *Rosaceae* the main fruit crops for human consumption.

In the ensuing chapters the first five of these families are discussed individually, followed by a briefer treatment of the *Rosaceae* and of the dozen or so other dicotyledonous families which contain crop plants of less importance in Britain.

# 3

# CRUCIFERAE

*General importance.* The family includes a small number of species of outstanding importance as 'root' crops. These are plants producing succulent leaves and young stems to provide bulk fodder for feeding green, or swollen storage organs suitable for folding-off or for temporary storage to provide winter fodder. Other forms of the same species are used for human consumption, and are important 'leaf' and 'root' vegetables. A few species are grown for their seeds, either as a source of edible oils (oil rapes) or for use as condiments (mustards). The family includes also a number of purely horticultural crops, and some important farm weeds.

*Botanical characters.* A family of herbaceous plants of temperate regions, with alternate, exstipulate simple leaves, often pinnately lobed. Inflorescence racemose, with few bracts. Flowers conspicuous, insect-pollinated, actinomorphic, of very characteristic and remarkably uniform structure. Floral formula K2+2, C4, A2+4, G(2). Sepals erect or spreading, inner pair sometimes pouched; petals usually with narrow erect basal claw and broader erect or spreading limb, forming, as seen from above, the characteristic cross from which the family derives its name. Stamens six (occasionally reduced to four or two), the outer pair shorter than the four inner. Pollen-sacs, four opening as two; nectaries present at base of filaments. Gynaecium superior, of two joined carpels, with a single short style and capitate or bilobed stigma. Ovules usually numerous, anatropous; placentation parietal, but the ovary divided into two chambers by the development of a false septum or *replum* as an outgrowth from the placentas. Pollination by insects, usually bees, sometimes by flies or small beetles, but self-pollination can occur. Fruit a specialized and characteristic form of capsule known as a *siliqua,* dehiscing by the carpel walls splitting longitudinally along the line of the placentas and becoming detached as separate *valves*, leaving the seeds temporarily attached to the replum. The siliqua is typically long and slender; the term *silicula* is used where the fruit is short and broad. Indehiscent or schizocarpic fruits occur in a few genera. Seeds are non-endospermic,

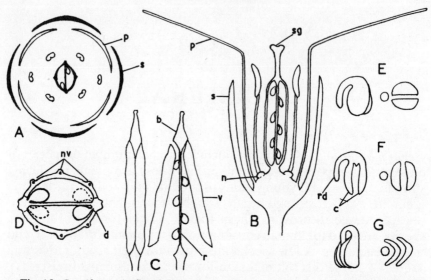

Fig. 10. *Cruciferae*. A, floral diagram. B, diagram of flower in vertical section. C, fruit (siliqua) entire and dehisced. D, fruit in transverse section. E, F, G, diagrams of side view and section of embryo to show arrangement in seed. E, accumbent. F, incumbent. G, conduplicate. *b*, beak. *c*, cotyledons. *d*, line of dehiscence. *n*, nectary. *nv*, nerve. *p*, petals. *r*, replum. *rd*, radicle. *s*, sepal. *sg*, stigma.

with thick cotyledons with mainly oily food reserves. In some species glucosides are present which are broken down by enzyme action to give strong-tasting sulphur compounds; such seeds are used as condiments, but are distasteful to stock and may be poisonous.

The arrangement of the embryo in the seed varies and may be of value as a systematic character. The embryo is bent through 180° in the region of the cotyledonary node, so that the radicle lies parallel to the cotyledons. If it lies adjacent to the edges of the two cotyledons it is described as *accumbent*; if along the mid-rib of one of them it is *incumbent*; in the latter case the cotyledons may be either flat or longitudinally folded around the radicle—*conduplicate*.

The great uniformity of flower-structure in the *Cruciferae* means that members of the family are readily recognized as such, but it makes the subdivision of the family difficult, and the distinctions between the tribes and genera are necessarily based on small characters. The family contains some 220 genera and 1 900 species, but of these only a very small number are of economic importance.

## BRASSICA

*Brassica* is by far the most important genus from the agricultural point of view. Plants of this genus are annual, biennial or perennial, with rather large pinnately-lobed or lyrate leaves, flowers with erect or somewhat spreading sepals and petals of some shade of yellow. In most species cross-pollination is normally necessary to ensure fertilization; genetic incompatibility factors prevent the satisfactory growth of pollen tubes following self-pollination. (Special techniques are used by plant breeders to overcome these incompatibility factors, thus enabling the inbred lines needed to produce F.1 hybrids to be developed.) Fruit a cylindrical or somewhat angular siliqua with globular seeds in one row in each loculus. Valves one-nerved, cotyledons conduplicate. Some forty species, of which the following are of agricultural importance:

### *Brassica oleracea* L.   Cabbages and Related Plants

Leaves all glabrous and glaucous; racemes extended, not corymbose, so that the unopened buds stand out above the open flowers. Flowers large (2·5 cm), sepals erect, petals pale yellow (very occasionally white). Outer stamens not much shorter than inner, filaments straight. Bracts sessile, narrowed to base, not clasping. Siliqua smooth with short conical beak. Seeds greyish-brown, 2 mm. Diploid chromosomes, 18.

*Wild form.* The species occurs wild as the perennial *sea cabbage*, found on the sea-coasts of southern and western Europe, including southern Britain. It bears somewhat fleshy sub-evergreen pinnately-lobed leaves in irregular rosettes on much-branched, rather woody stems. Flowers, fruits and seeds are similar to those of the cultivated forms.

*Cultivated forms.* No precise information is available on the place and time at which the species are brought into cultivation, but it may be assumed that the wild form was early used for human food, and probably more tender and less bitter types selected for cultivation; certainly forms of the species have been grown in the Mediterranean area for well over 2 000 years. Almost all the forms now in cultivation are biennial; they fall into a number of well-marked groups according to the particular part of the plant in which increased size and succulence have developed. The different groups represent enlargement of all parts of the plant other than root, fruit and seed. It should be emphasized that all forms are very similar in flower and fruit, and that they cannot be distinguished one from another in the seed stage.

GROUP 1. KALES*

In the kales, which may be regarded as the nearest to the wild form, it is the young stems and loosely arranged leaves which form the enlarged edible part. The larger-growing, higher-yielding type are the agricultural kales; the smaller forms of similar habit are garden vegetables. All forms are used in the vegetative stage, usually during the winter; if they are allowed to stand they produce inflorescences in the spring and die after ripening seed in the succeeding summer.

*Thousand-head kale.* A tall plant (up to 1·2 m) with rather slender woody main stem bearing on its upper part only leafy succulent branch shoots with plane or only slightly crinkled leaves. Moderately frost-hardy, and available, from a late spring sowing, for use during autumn and winter; development of inflorescences with consequent decrease in feeding value takes place in spring. Yields may be up to 55 tonnes per hectare with about 13% dry matter containing some 18% protein and with a D-value of about 67. Dwarf thousand-head kales, including Canson, are hardy, generally later flowering, with usually a good growth of new shoots late in winter, and are thus particularly suitable for feeding in February and March. They are however lower yielding, giving some 50 t/ha with similar dry matter content, but with protein up to 20% and D-value up to 70.

*Marrow-stem kale (Chou Moellier).* This closely resembles thousand-head kale, but the stem is much thickened throughout the greater part of its length. This thickened stem, up to 10 cm in diameter, is edible, and shows in transverse section a very large succulent pith surrounded by a narrow ring of small vascular bundles. The stem-structure thus resembles that of kohlrabi (p. 66), from which marrow-stem kale is perhaps derived by crossing with thousand-head; there is however no definite evidence on the origin of marrow-stem kale, which was first recorded in France early in the nineteenth century. Yields may be 50 to 70 t/ha at about 12% dry matter with D-value 71 and protein 18% on the whole crop; the feeding value of the thickened stem (*c*. 12% protein) is less than that of the leaves (*c*. 20%).

Marrow-stem kale is rather less hardy than thousand-head, and is usually fed during autumn and early winter. Green and purple-skinned varieties exist, of which the former is much the more common. Both thousand-head and marrow-stem kale are very valuable

* 'Kale' is a habit description, and plants referred to as kales are also found in the species *B. napus* (p. 72).

crops providing large yields of high-protein fodder available for winter feeding, and as they can be grown without hand thinning, and fed off *in situ,* are cheaper to grow than the majority of 'root' crops.

*Cultivars.* Since the late 1960s commercial stocks of the kales have tended to be replaced by bred cultivars. Maris Kestrel, a hardy hybrid marrow-stem kale, has been particularly successful. It resembles a short-stemmed marrow-stem kale in appearance but produces higher dry matter yields; it has a higher leaf-to-stem ratio and this, coupled with a more succulent stem, gives it a much higher overall D-value. It is as hardy as thousand-head kale and can be used throughout the kale season. The shorter stem makes it more suited to strip grazing. It was originally developed as a double cross hybrid, in which one of the four parent lines was probably a marrow-stem kale × hardy cabbage hybrid; during the 1970s it was further developed as a triple cross hybrid. Other named cultivars have been produced in Britain and on the continent and show improvements in yield and quality when compared with standard commercial stocks. Work at the Plant Breeding Institute at Cambridge, where Maris Kestrel was produced, suggests that further improvement may be possible using material derived from hybrids of kales with brussels sprouts, the latter adding hardiness and improved digestibility.

*Horticultural kales (Borecole).* A large range of *Brassica oleracea* kales were formly grown as private garden vegetables. Picked shoots (greens) wilt rather readily, and do not travel well, and kales for human consumption are therefore not important commercially. Curly kale, with densely crisped leaves, is still used, and Pentland Brig, a bred cultivar of this type, but producing edible axillary shoots in spring, was introduced during the 1970s. Variegated kales, with white, pink or purple variegation of the leaves, are available for ornamental purposes.

GROUP 2. CABBAGES

In the cabbages, the main feature is the head of closely-packed leaves, which may be regarded as an immensely enlarged terminal bud in the vegetative stage. This stage is usually reached in autumn, winter or early spring, according to variety and time of sowing; if the plant is allowed to stand, the growing point enters the reproductive phase and the upper internodes lengthen, giving a long branched inflorescence in the succeeding summer. In the very dense-headed cabbage the

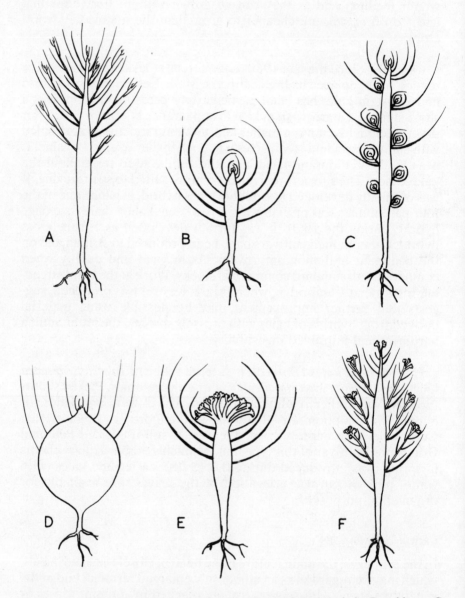

Fig. 11. *Brassica oleracea,* diagrams of cultivated forms. A, kale. B, cabbage. C, brussels sprouts. D, kohlrabi. E, cauliflower. F, sprouting broccoli.

closely-packed lower leaves may, however, interfere with the development of the flowering stems, and for seed production it is often necessary either to slash the leaves vertically to allow the inflorescence to emerge, or to cut the head and allow flowering branches to develop from the lower part of the stem. Dense-headed cabbages may perhaps have been evolved in Germany, where they are recorded from the twelfth century; earlier cabbages appear to have been loose-headed non-hearting forms. Great variation exists in the size and shape of the head in different types of cabbage; the larger forms are used for fodder, the smaller for human consumption.

*Cattle cabbages.* Large-headed cabbages, usually sown in special seed-beds and planted out at about 1 m square; slow maturing. The main form is the drumhead or flatpoll cabbage with broad flattened heads, of which numerous varieties exist. Yields may be up to perhaps 100 t/ha, but are more usually about 60 t/ha, with about 10% dry matter containing some 22% protein and with a D-value of about 71; the cost of growing is however usually greater than for kale.

*Cabbages for human consumption.* Smaller forms, widely grown on a field or garden scale, and utilized as vegetables, include very many varieties, varying in size, in shape (mainly with round or pointed heads) and in time of maturity (from summer to late winter from spring sowings, and spring from late summer sowing). The development of numerous F.1 hybrid cultivars which show great uniformity in size, shape and time of maturity, coupled with improvements in husbandry techniques which enable crops to be direct drilled, has led to a high degree of mechanization of the cabbage crop, enabling single destructive harvests to be made of cabbages sufficiently uniform to be suitable for mechanical handling and prepacking for sale. Some of the tight round-headed cabbages may also be stored under suitable conditions for release on to the market over a considerable period.

Distinct forms are red pickling cabbage, usually large slow-maturing forms with anthocyanin pigments in the leaves masking the chlorophyll, and savoys, similar in habit to round-headed cabbage but with thick puckered leaves, hardy, perhaps Italian in origin, recorded from the sixteenth century. Coleworts (collards), hardy loose- or non-hearting small cabbages, and couve-tronchuda (Portugal cabbage) with succulent white enlarged petioles and midribs, are now largely obsolete in Britain.

GROUP 3. KOHLRABI

Kohlrabi is a very distinct form in which the stem is enlarged to form a globular or fusiform 'bulb'. The mature plant at the end of the first year shows a short length of slender, woody stem surmounted by the 'bulb' which shows the widely-spaced leaf-scars on its lower part; leaves are present on the upper part of the 'bulb' and at the top is a very short leaf-bearing 'neck' which elongates only in the second year of growth.

The kohlrabi 'bulb' thus differs from that of the swede (see p. 73) in consisting of stem only. Its anatomical structure is also different, in that a transverse section shows a narrow ring of small vascular bundles with normal lignified xylem surrounding a much-enlarged pith, which forms the bulk of the 'bulb'. In the pith are slender branching strands of vacular tissue with some lignified cells.

Since the 'bulb', which stores well, forms the greater part of the yield, kohlrabi is comparable in its agricultural use to turnips and swedes; it may be used to replace these under conditions where its greater drought-resistance or its ability to withstand transplanting is

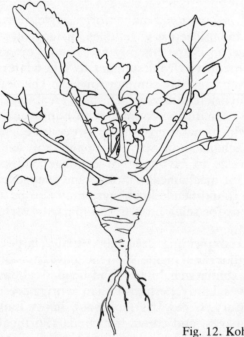

Fig. 12. Kohlrabi, young plant, × ¼.

an advantage. It is used also to some extent as a garden vegetable. Yields may be *c*. 50 t/ha.

Little variation is shown in different stocks, but green-skinned and purple-skinned varieties exist.

The remaining groups within the species contain comparatively low-yielding forms grown for human consumption only.

### GROUP 4. BRUSSELS SPROUTS

Comparable to cabbages, but here it is the axillary buds, not the terminal, which are eaten; the axillary buds forming in the vegetative stage a compact globular mass of tightly-packed leaves borne on a very short stem. A form of relatively recent origin, first clearly recorded in Belgium in the mid-eighteenth century, although some thirteenth-century records may refer to this form. The 'sprouts' are available for use in autumn and winter only; numerous cultivars exist, varying in length of stem and size of sprouts and, within the relatively limited period, in date of maturity. Sprouts may also vary in texture (rough or smooth); in solidity (loose or solid); in shape (round or oval); in colour (dark or light green, purple forms also exist); and in the closeness of spacing along the stem.

Traditional open-pollinated cultivars mature sprouts from below upwards over a period of some weeks and are normally picked over by hand several times as the sprouts mature. Much of the commercial crop is now machine-picked at a single harvest, and for this special cultivars (mainly F.1 hybrids) and special cultivation techniques have been developed. The cultivars produce a uniform crop of relatively small solid sprouts which are resistant to growing out as leafy shoots; harvest can thus be delayed until the sprouts towards the top of the stem are sufficiently well developed. Crops for single-harvest treatment are grown at close spacing and for early and mid-season crops 'stopping' (removal of terminal bud) is commonly practised; this is usually done some three to six weeks before the expected harvest, the optimum time being when the lower sprouts are about 12·5 mm in diameter. The effect of close spacing and stopping is to produce greater uniformity in sprout development along the stem.

### GROUP 5. CAULIFLOWER

Here the terminal bud is enlarged in the flowering stage, so that the head consists of short, much-swollen inflorescence branches bearing a compact mass of tightly appressed flower buds in an early stage of differentiation, the whole white or very pale yellow inflorescence (curd) being surrounded and protected by the upper leaves. The whole inflorescence, consisting mainly of meristematic cells, has a very high protein content. Cauliflowers are not recorded before the middle ages, and were probably introduced into Italy from the eastern Mediterranean area in the late fifteenth century; some authorities regard them as having been derived from *Brassica cretica* Lam., indigenous to that area, rather than from *B. oleracea* in the strict sense. They reached north-western Europe early in the seventeenth century.

Forms are now available which come to the cutting stage at almost all periods of the year. Purple, green and yellow types exist, but are of no commercial importance; there is a very strong market preference for white curds, hemispherical, compact and free from separated flower buds and from green bracts. The greater part of the crop is marketed fresh, graded and carefully packed, but some is deep-frozen as 'cauliflower-sprigs' (separate curd branches), and some used for pickles.

*Cultivars*. A large number of cultivars exist and these may be grouped according to heading date. This is controlled by day-length, and in the case of winter-heading and winter-hardy forms by a cold requirement for vernalization, effective only when the plant has reached a certain minimum size. The range of cultivars is such that cauliflowers may be produced all the year round, although it is only in the milder areas that the winter heading forms (sometimes referred to by the grower as broccoli) can be safely grown. Summer cauliflowers are commonly raised under glass and planted out after the soil begins to warm up in spring. Current open-pollinated cultivars, although showing greater uniformity of size, shape and period of maturity than the older mass-selected types, mature over a period of several weeks. Non-selective single harvesting is only practicable where F.1 hybrid cultivars are available; technical difficulties in the production of these are being resolved.

*Nine-star perennial broccoli* is a distinct form occasionally grown in gardens, with a branching stem producing several heads and capable of persisting and heading for several years.

GROUP 6. SPROUTING BROCCOLI

In these the axillary buds form short, branching inflorescence-bearing shoots; it is these shoots, each terminating in a very small 'curd' comparable in stage to that of cauliflower, which are eaten. The majority of forms mature in late spring, but early forms exist. The small curd may be purple, white, or green (calabrese).

Until recently the crop was little grown commercially, since, as with the horticultural kales, the picked shoots do not travel well. The introduction of new marketing methods has altered this, the shoots ('broccoli spears') being deep-frozen or packed in polythene to prevent wilting. Calabrese forms, maturing in late summer and early autumn, and producing small terminal curds followed by a succession of axillary curds, are used. With F.1 hybrid calabrese cultivars, direct drilled at close spacing to suppress development of axillary shoots, non-selective single harvesting by machine is possible; with this technique only the primary heads are used, and the plant is in effect being treated as a miniature cauliflower, rather than a sprouting broccoli in the usual sense.

### *Brassica napus* L.    Swedes and Related Plants

The leaves of the mature plant of *Brassica napus* are glabrous and glaucous like those of *B. oleracea,* but the young plant shows leaves which are somewhat hairy. The flower is smaller than that of *B. oleracea*; sepals somewhat spreading, petals either bright yellow or buff; the filaments of outer stamens shorter and more curved. The inflorescence is more corymbose at the flowering stage, so that the unopened buds tend to be on the same level as the open flowers. Leaves on inflorescence with cordate partly clasping base. The siliqua is very similar to that of *B. oleracea,* but the seeds are purplish-black, without the brownish tinge seen in that species; size varying with the form, but usually slightly larger (2·2 mm) in rapes. Diploid chromosomes, 38. Largely self-pollinated, with crossing in swedes usually 33% or less.

*Brassica napus* is not known as a wild plant, and can be regarded as a synthetic species derived from *B. oleracea* and *B. rapa* (see p. 79). The cultivated forms show considerable variation, and may be grouped as follows:

GROUP 1. RAPES (SWEDE-LIKE RAPES, COLESEED)

Rapes are forms with rather slender, usually branching stems, with rather small leaves and unthickened root. They are grown for two distinct purposes, (1) as an oil-seed crop, (2) as a leafy forage crop.

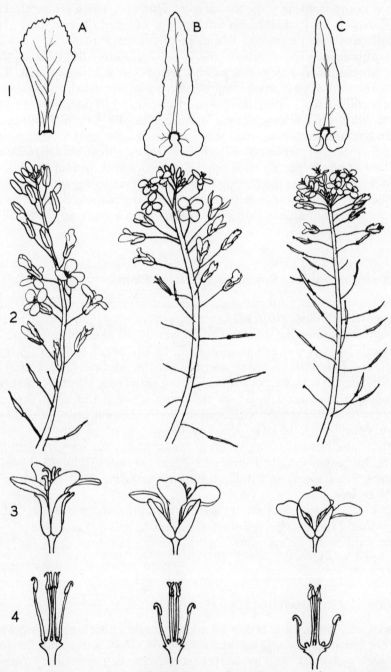

Fig. 13. 1, bract, × $\frac{2}{5}$. 2, inflorescence, × $\frac{2}{5}$. 3, flower, × $1\frac{1}{4}$. 4, diagram of arrangement of stamens, × $1\frac{1}{2}$. A, *Brassica oleracea*. B, *B. napus*. C, *B. rapa*.

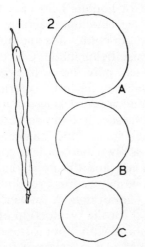

Fig. 14. 1, pod of *Brassica oleracea* × 1. 2, outlines of seeds of B, *B. oleracea*, A, *B. napus*, C, *B. rapa*, × 10.

*Oil rapes.* Swede-like oil rapes, rather extensively grown in various European countries for oil (rape oil, colza oil, formerly used for burning, and at the present day for lubrication and for margarine manufacture), exist in two forms, the biennial winter oil rape sown from mid-August to the first week in September and harvested the following July, and the quicker growing summer oil rape sown in late March and early April and harvested from mid-August to September.

The importance of oil rape in British agriculture developed in the late 1960s following world increases in oil seed demand. Interest was at first mainly in summer oil rape but tended to be transferred to winter oil rape in the late 1970s. These give higher yields of better quality seed, some 2 t/ha at 46% oil content, compared with about 1·5 t/ha at 44% oil for summmer oil rape. They have however the disadvantage that their early sowing date makes the choice of preceding crop more difficult.

A problem with rape oil, which ranks fifth in importance amongst vegetable oils on a world scale, has been the high erucic acid content (45–54% of the total fatty acid content). Erucic acid has been associated with the production of fatty deposits in the arteries of experimental animals and is therefore considered undesirable in the human diet. Cultivars with erucic acid content in the range of 0–11% of total fatty acid content have been developed, and it is these forms which are likely to be grown in the future, as E.E.C. policy is to limit the amount of erucic acid permitted in rape oil for human consumption. For some industrial uses however a high erucic acid content is an advantage. Plant breeders are also trying to reduce the level of linolenic acid (5–15% of total fatty acids) as this may be associated with tainting, and to increase the level of linoleic acid (11–31%), desirable in oils for culinary use and margarine manufacture.

Attempts to reduce the glucosinolate (thioglucoside) content of the residues left after oil extraction are also being made. Glucosinolates, although harmless in themselves, may be converted to thiocyanates and oxazolidine-thione which are harmful to farm animals, especially to pigs and poultry. Modern processing techniques have largely overcome this problem and the seed residues left after oil extraction form a valuable protein-rich stock food marketed as rape-meal or rape-cake. Cultivars with a low glucosinolate content would however still have some advantages, and the Polish cultivar Bronowski, which has this character, is being used as breeding material at the Plant Breeding Institute at Cambridge. A further plant breeding technique, which has been used for the production of new oil rape cultivars, and which might be utilized with other forms of the species, is their *de novo* synthesis by artificial crossing of suitable forms of *Brassica oleracea* and *B. rapa* followed by chromosome doubling with colchicine to give the fertile amphidiploid.

*Forage rapes*. Rape is commonly grown in Britain as a quick-growing forage crop for grazing off; the forms employed are similar to the winter oil rapes in being biennial. The earlier classification into two distinct forms, giant and dwarf, is no longer tenable, as the cultivars now available form a more or less continuous series from those with erect single stems of up to 0·8 m tall down to the prostrate more spreading forms with stems less than 0·5 m tall. Few of the very short-stemmed types are grown, and none was listed in the 1977/8 N.I.A.B. leaflet, *Recommended Varieties of Green Fodder Crops,* which gives a range of from 0·64 m to 0·81 m for stem height. Average yields in trials were some 42 t/ha with some 12% of dry matter with a D-value of about 70 and containing some 22% protein. D-values and protein content tend to be slightly higher in the dwarfer cultivars, including the club-root resistant but mildew susceptible Nevin, but the yield of digestible organic dry matter is lower than from the taller cultivars. Forage rapes will not normally compare in yield with the *B. oleracea* kales, and are employed mainly where speed of growth is the main consideration, or where a 'nurse-crop' for grass and clover seeds is required.

GROUP 2. SWEDE-LIKE KALES

Forms of the species with rather stout stems and large leaves are referred to as kales, from their resemblance in habit to the *B. oleracea* kales. The yield per acre is much lower than from the latter, but they provide feed in late spring when few other forage crops are available. Two forms are grown for stock feed, *rape kale (sheep kale),* with stout, little-branched stems, and *hungry-gap kale,* with rather more slender, more branching stems, maturing later. *Asparagus kale,* formerly grown as a garden vegetable for its leafy axillary shoots in the spring, is now no longer maintained.

GROUP 3. SWEDES (SWEDE TURNIPS, SWEDISH TURNIPS, RUTABAGAS (U.S.A.))

In swedes a swollen 'root' is produced in the first season, which can be fed off in autumn or lifted and stored for winter use. The swollen region forms a massive structure partly above ground; it consists of the upper part of the true root ( it must be emphasized that it is only a small part of the root system which is swollen, and that the lower part of the root forms an extensive branching system extending to a depth of 1·5 m and spreading laterally to a distance of 0·6–0·75 m), the hypocotyl, and the lower part of the true stem. The greater part of the stem forms a conspicuous, leaf-bearing 'neck', which elongates in the second year to form a tall, branched inflorescence.

The swollen part of the plant consists mainly of parenchyma, arising not as pith (as in kohlrabi), but as unlignified secondary xylem. A transverse section through the lower part shows a slender diarch primary xylem, surrounded by a very broad ring of secondary xylem composed mainly of thin-walled parenchymatous cells, with only occasional scattered lignified cells which are formed along radii separated by rays of parenchyma tissue. Along the same radii as the lignified xylem elements are associated parenchyma cells which divide to form irregular secondary cambia giving rise to small clusters of tertiary phloem cells; these are presumably of significance in the translocation of food reserve materials. The secondary phloem, outside the cambium, also consists of thin-walled cells; it is surrounded on the outside by a thin layer of cork cells forming the 'skin' of the root. In the upper part of the root a small pith is present and the neck shows typical stem structure with a rather large pith surrounded by a narrow ring of xylem, here mainly lignified.

Cambial activity normally stops during the winter; in the following spring further and much more heavily lignified secondary xylem is produced, and food material stored in the parenchyma is withdrawn to provide for the rapid growth of the stem. The 'root' thus decreases greatly in palatability and feeding value if it is not used before growth recommences.

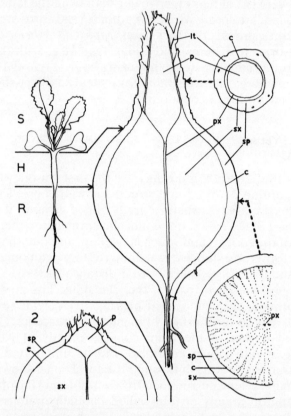

Fig. 15. 1, swede; diagram of seedling and of mature 'root' in longitudinal and transverse section. 2, turnip, longitudinal section of upper part of 'root' to show shorter 'neck'. S, stem region; H, hypocotyl; R, root. *c,* cambium. *lt,* leaf-trace. *p,* pith. *px,* primary xylem. *sp,* secondary phloem and cortex. *sx,* secondary xylem.

The mature 'root' of different forms of swedes varies in flesh colour, which may be yellow or white (flower colour is correlated with this, white swedes having yellow flowers, like those of rape, while the yellow-fleshed swedes have buff flowers), in 'skin' colour, and in shape. Yellow-fleshed forms are much more commonly grown than white; purple, green and bronze tops, due to the development of anthocyanin pigment or chlorophyll or both in the outer cells of that part of the 'root' which is exposed to light, are found associated with both types of flesh colour. Shapes are usually described as globe or tankard, but may be intermediate between these two. There appears to be no constant correlation between this colour and shape grouping and agricultural value, but green and dark-purple skinned forms are mainly frost-hardy, while the light-purple forms are readily damaged by frost. Within each of the more popular groups a number of named varieties exist; these differ from one another only in minor or in physiological characters. Swedes being partially cross-pollinated, variation will be found both from stock to stock of the same named variety, and from plant to plant within the same stock.

Swedes are normally grown only as a wide-drilled, thinned crop, and are therefore an expensive crop to grow. The yield, in the region of 60 t/ha, with a dry matter of 9–10%, giving a dry matter yield of around 5·5–6 t/ha, does not, at least in the south of England, compare favourably with that of mangels or fodder-beet. The main advantages of swedes are:

(1) Frost-hardiness, which enables them to be used for late autumn folding; for early folding the quicker-growing common turnips can be employed, and there appears to be little place for those swede varieties which show poor frost resistance.

(2) Their saleability, in some areas, for human consumption as a winter vegetable. Purple-top yellow-fleshed globe varieties which produce uniform roots of relatively small size are preferred for this purpose (e.g. Acme).

## *Brassica rapa* L.*    Turnips and Turnip Rapes

*B. rapa* is distinguished from the two preceding species by its bright-green, not glaucous, densely rough-hairy lower leaves. The inflorescence forms a corymb at the flowering stage, with open flowers standing well above the unopened younger buds. Leaves on the inflorescence are usually glabrous and often somewhat glaucous, but with more deeply cordate and more markedly clasping bases than those of *B. napus*. Flowers smaller than in that species, the outer stamens with still shorter curved filaments. Siliqua closely resembling that of the two previous species; seed smaller and more variable in colour; some seeds are purple-black, but others are brighter red-purple, so that a bulk sample always has a paler and redder look than seed of *B. napus*.

* This name applied originally only to the turnips; wild turnip and turnip rapes were named *B. campestris* L. The earliest treatment recognizing that all these belonged to one single species was that of Metzger (1833) who used *B. rapa* L. for the combined species, and this therefore appears to be the valid name. (*B. campestris* is still used for rapes in the 1973 British national list, following E.E.C. practice, see p. 83.)

*Wild form.* A biennial form with a slender tap-root is found in Britain (possibly introduced) as a locally common riverside plant; an annual form is a charlock-like weed in some areas.

### Cultivated forms

The range of cultivated forms is less wide than in the swede species, and until the mid-1970s, when some turnip forage rapes came into trials, included only oil rapes and bulbous-rooted turnips.

*Oil rapes.* The two forms of turnip-like oil rape, winter and summer, correspond to the two swede-like oil rapes and are used in the same way; their yield is lower, but the plants are hardier and will tolerate poorer soils. They therefore replace the swede rapes under these conditions; it does not appear likely that they would be of great value in Britain. They are however grown in Canada, parts of Scandinavia and elsewhere where the conditions are less suitable for swede oil rapes. The average oil content is less than for swede oil rapes, down to about 40% in summer cultivars. Low erucic acid forms have been produced.

*Forage rapes.* Turnip rapes for forage use were developed in the mid-1970s, when the cultivars Appin and Ballater were produced by the Scottish Plant Breeding Station. These are tetraploids derived from crosses of Tigra, a club-root resistant stubble turnip (see below) with an oriental salad vegetable belonging to the Asiatic subspecies *nipposinica* (Bail.) Olss. These turnip forage rapes are non-bulbing forms with distinctive dissected leaves; they have the ability to produce regrowth after grazing. Cultivars of similar type, but with larger less dissected leaves, have been produced in Holland.

*Turnips* (common turnips). The turnips are biennial forms of this species, with the upper part of the root, the hypocotyl, and the lower part of the stem swollen by the development of extensive, little lignified secondary xylem. They thus correspond very closely in structure to swedes; the main difference is that the 'neck' in turnips is very short, the pith having the form of an obtuse cone, instead of a long cylinder, as in swedes. Turnips show a much greater range of 'root' shape than swedes, the length-breadth ratio varying from about 6 in extreme long forms to 0·5 in the flat forcing types; the agricultural turnips are, however, almost all globe or semi-tankard forms. Turnips are in general quicker maturing than swedes, and better adapted to poor conditions, but are of lower feeding value.

Many cultivars are used for human consumption, and turnips are classed as vegetables in the National List.

*(1) White-fleshed turnips.* Quick-maturing forms, with low dry-matter content (7–8%), normally sown in summer for folding off in autumn. They are not frost-hardy and do not store well, and must be used as soon as the 'root' has reached its maximum size; early-sown plants may produce elongated stems and inflorescence in the same year (bolt); when this occurs a rapid drying-out of the xylem-parenchyma takes place and the tissue becomes 'pithy' and inedible. White-fleshed turnips may be white, green, red, or bronze topped. Flowers bright yellow.

Stubble turnips are quick-growing large-leaved turnips grown to provide autumn forage. Some typical British flat or globe forms are used in this way, but cultivars introduced from Holland are more commonly employed; these have large, often more or less entire leaves, and cylindrical roots. The swollen root loses palatability rapidly, and is often completely neglected by stock unless the crop is utilized while the root is still actively growing. Some very vigorous autotetraploid cultivars have been developed in Holland.

*(2) Yellow-fleshed turnips.* (Scotch Yellows.) Yellow-fleshed turnips have a higher dry-matter content (8–9%), are slower-maturing and store better than white-fleshed turnips; they are normally sown earlier and used later, and are thus intermediate in agricultural use (but not in botanical characters) between white-fleshed turnips and swedes. The majority of varieties used are green or purple topped; the flowers are buff. The remarks made regarding named varieties in swedes (p. 75) will apply equally to turnip varieties.

### *Brassica nigra* (L.) Koch. (*Sinapis nigra* L.).   Black Mustard

An annual, with dark-green, sparsely-hairy lyrate leaves, base of inflorescence-leaves narrowed into a stalk; flowers small, bright yellow, with spreading sepals; siliquas smooth, more or less four-sided, short, with very short beak, erect on short pedicels. Seeds small, brownish or reddish to black, conspicuously pitted.

Black mustard occurs as a wild plant in southern England, occasionally as an arable weed. It was grown to a limited extent in eastern England as a crop plant for the seed, which was used in the production of table mustard (cf. also *Sinapis alba*, p. 84). Its unpalatable nature and low vegetative yield made it useless as forage, and as a seed crop it suffered from the disadvantages of ready pre-harvest seed shedding, the ability of the seed to remain dormant in the soil, and thus a tendency to persist as a weed in succeeding crops. Since the 1950s it has been almost completely replaced as a crop by forms of *B. juncea* (see below).

The seeds contain the glucoside sinigrin (potassium myronate, potassium allyl glucosinolate), which, when the ground-up seed is mixed with water, is hydrolyzed by the enzyme myrosinase to give the volatile, pungent, mustard-smelling allyl isothiocyanate. An oil, somewhat similiar to rape oil, can be extracted from the seed, but the residues are unpalatable and may be injurious, and should not be used in oil-cake manufacture.

### *Brassica juncea* (L.) Czern.   **Brown Mustard,** Chinese Mustard, Indian Mustard

A self-pollinating annual of Asiatic origin. Leaves usually slightly glaucous and hispid, somewhat swede-like; inflorescence and flowers similar to turnip, but with linear-lanceolate bracts with narrow stalks not clasping the stem. Siliquas cylindrical, semi-erect; seeds small, brown or yellow. Diploid chromosomes 36; the species is apparently an allopolyploid derived from *B. nigra* and *B. rapa* or a similar species. Heat-resistant; various leafy forms are grown as vegetables in Asia and elsewhere in climates too hot for the satisfactory use of temperate species, and seed-producing forms for oil production under similar conditions.

In Britain a small number of specially developed cultivars of the seed-producing type are grown on a limited scale under contract for the production of 'black mustard'. First introduced in 1952, this brown mustard rapidly and almost completely replaced the true black mustard (*B. nigra*) formerly grown for this purpose. Brown mustard is a better agricultural crop, more compact, less prone to shedding, more suitable for combine harvesting, and the product is very similar. Cultivars are largely pure lines; those with yellow seeds are preferred as having a smaller proportion of testa than the brown-seeded.

*Other Brassica species.* Some other species are cultivated; these include *B. pekinensis* Rupr. and *B. chinensis* L., both with a diploid chromosome number of 20 and both known as Chinese cabbage. These are widely grown in Asia as vegetables; loose-leafed and dense-headed cabbage forms exist. Occasionally grown, mainly on a garden scale, in Britain and sometimes marketed as 'Chinese leaves'.

## Inter-relations of Brassica species

Three primary groups of species exist with differing chromosome numbers:

$$n = 9 \quad B. \; oleracea$$
$$n = 10 \quad B. \; rapa$$
$$B. \; pekinensis$$
$$n = 8 \quad B. \; nigra$$

These different genomes are conveniently indicated by letters; the ten chromosomes of *B. rapa* are referred to as *a,* the eight of *B. nigra* as *b,* and the nine of *B. oleracea* as *c.*

Crossing can take place between members of the same groups, and occasionally between members of different groups; in this last case the hybrids will be sterile. If, however, chromosome doubling takes place, then pairing of chromosomes derived from the same species can take place at meiosis, and these doubled hybrids (amphidiploids) are fully fertile and behave as a new synthetic species. In this way *B. napus* (rape and swedes) has arisen, by natural crossing (in itself very rare) between *B. oleracea* and *B. rapa,* followed by chromosome doubling to give thirty-eight chromosomes *aacc.* The cross apparently occurred in central Europe some time during the seventeenth century or earlier, and perhaps on a number of occasions; it has been artificially repeated frequently, and even used for the synthesis of new cultivars of *B. napus.* Two other cultivated species, *B. juncea aabb,* and *B. carinata* (Abyssinian cabbage) *bbcc,* have apparently arisen in the same way.

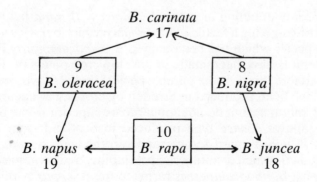

### Crossing in Brassica species

All *Brassica* species are normally cross-pollinated, and all forms belonging to one species cross readily. Thus all the different forms of *B. oleracea* will cross-pollinate one another, and the utmost care must therefore be taken when growing crops for seed, to ensure that crossing between different forms cannot take place. For instance, if a crop of cauliflower were being grown for seed, and, say, marrow-stem kale plants were allowed to flower in the same or an adjoining field; then, if flowers of both forms were open at the same time, cross-pollination would be likely, and the seed saved from the cauliflower crop would contain a proportion of worthless cauliflower-kale hybrids. Similarly, in the species *B. napus,* the different forms of swede, rape and swede-kale will all cross together.

In addition to this very ready crossing between different forms of the same species, crosses may occur between different species with sufficient frequency to make isolation desirable. *B. oleracea* can be assumed, for all practical purposes, not to cross with any other species, but the group of species with ten chromosomes in the haploid state, *B. rapa, B. pekinensis* and the amphidiploid 'synthetic species' *B. napus* and *B. juncea,* derived from them, will all cross together and should therefore be isolated from one another when grown for seed. Two only of these species—namely, turnip and swede (including swede-rapes and kales)—are common enough in Britain to be considered; here, crossing is more likely to be found where the swede is the seed parent than where the reverse is true. The swede-turnip hybrids *aac* so produced are normally sterile and sometimes show irregular swellings on the roots which have been described as 'hybridization nodules'. Crossing between *B. juncea* and *B. nigra* is also possible.

Chromosome doubling in sterile *B. napus* × *B. rapa* hybrids can restore fertility, giving a fertile hybrid *aaaacc* which behaves as a new synthetic species which has been named *B. napocampestris* Fran. et Wing. Attempts have been made to develop crop plants of this type by using suitable forms of the parent species; thus *B. napocampestris* oil rapes have been produced in Sweden by crossing swede oil rapes with turnip oil rapes, but do not appear to be superior to the parents. A forage rape of *aaaacc* type produced in Scotland from a cross between a *B. napus* dwarf forage rape and the leafy *B. rapa* subsp. *nipposinica* had some advantages in palatability and protein content. Back-crossing *B. napocampestris* forms to the *B. napus* parent gives *aaacc* pentaploids from which new and improved types of *B. napus* might be recovered if the extra *a* chromosomes were elimated in later generations. Similarly, back-crossing to the other parent, *B. oleracea,* might be of value, but this appears to be a very difficult cross to make.

### Isolation of seed crops

For practical purposes a distance of over 200 m is regarded as providing satisfactory isolation between two cross-pollinating crops, and it is normally considered necessary, therefore, that seed crops of different forms of one species, or of different varieties within one form, should be separated by at least this distance. This figure of 200 m is, however, a purely arbitrary one; the likelihood of cross-pollination falls off very rapidly in the first 100 m, but the further decrease with distance is very slow, and while 200 m is well beyond the distance at which any bulk crossing will take place, it is still not sufficient to exclude entirely occasional cross-pollination by chance far-flying insects or even by the small but appreciable amount of wind-borne pollen. Such chance cross-pollination will, however, result in only a negligible proportion of hybrids in the seed, and from a practical point of view can be disregarded.

The current Seeds Regulations for forage and oil seed *Brassica* crops stipulate a minimum isolation distance of 200 m for certified seed and 400 m for basic seed.

Even where different cultivars are flowering in close proximity there may be relatively little cross-pollination between them, since bees working the flowers may confine their visits to one cultivar or line and almost completely ignore others; this mode of behaviour of bees can in fact cause difficulties where the crossing of two interplanted inbred lines is required for the production of F.1 hybrid seed.

### Nomenclature of groups within the Brassica species

The different kinds of crop plants within the *Brassica* species have been referred to here by their common English names, e.g. cabbage, brussels sprouts, etc. This is quite satisfactory for use in this country, but the adherence of Britain to the E.E.C. makes the use of internationally acceptable names necessary for Community purposes. The only international names for the different kinds of crop plants are Latin botanical names, and these therefore have to be used. The standard botanical system of nomenclature is however, as has been stated earlier (p. 54), designed for wild plants and is not well suited to the distinguishing of different kinds of crops within a species, least of all when the array of these is as extensive as it is in the *Brassica*

*3. Cruciferae*

## Table 1.   Nomenclature of *Brassica* crops

| 1<br>ENGLISH NAME | 2<br>E.E.C. NAME | 3<br>MANSFELD |
|---|---|---|
| ***Brassica oleracea* L.** | | |
| 1 000-head kale | convar. *acephala* DC. | convar. *oleracea* var. *mille-capitata* (Lev.) Helm. |
| Marrow-stem kale | | convar. *acephala* (DC.) Alef. var. *medullosa* Thell. |
| Curly kale | var. *acephala* DC. subvar. *laciniata* L. | convar. *acephala* (DC.) Alef. var. *sabellica* L. |
| Kohlrabi | var. *gongylodes* L. | convar. *acephala* (DC.) Alef. var. *gongylodes* L. |
| Cabbage | var. *capitata* L. f. *alba* DC. et f. *rubra* (L.) Thell. | convar. *capitata* (L.) Alef. var. *capitata* |
| Savoy | var. *bullata* DC. et var. *subauda* L. | convar. *capitata* (L.) Alef. var. *subauda* L. |
| Brussels sprouts | var. *bullata* DC. subvar. *gemmifera* DC. | convar. *oleracea* var. *gemmifera* DC. |
| Cauliflower | convar. *botrytis* (L.) Alef. var. *botrytis* | convar. *botrytis* (L.) Alef. var. *botrytis* |
| Sprouting broccoli | convar. *botrytis* (L.) Alef. var. *italica* Plenck | convar. *botrytis* (L.) Alef. var. *italica* Plenck |
| ***Brassica napus* L.** | | |
| Swede | var. *napobrassica* (L.) Peterm. | var. *napobrassica* (L.) Rchb. |
| Rape kale<br>Forage rape<br>Winter oil rape<br>Summer oil rape | subsp. *oleifera* (Metzg.) Sinsk. | var. *napus* f. *biennis* (Schübl. et Mart.) Thell.<br>var. *napus* f. *annua* (Schübl. et Mart.) Thell. |
| ***Brassica rapa* L.** | | |
| Turnip | *B. rapa* L. var. *rapa* (L.) Thell. | *B. rapa* L. var. *rapa* |
| Winter turnip rape<br><br>Summer turnip rape | *B. campestris* L. subsp. *oleifera* (Metzg.) Sinsk. | *B. rapa* L. var. *sylvestris* (Lam.) Briggs f. *autumnalis* (DC.) Mansf.<br>*B. rapa* L. var. *sylvestris* (Lam.) Briggs f. *praecox* (DC.) Mansf. |

subsp. = subspecies    convar. = convarietas    var. = varietas
subvar. = subvarietas    f. =forma

species. There is moreover no real consensus of opinion on which of the very numerous names which have been proposed are the valid ones. Those selected by the E.E.C. are the ones that in practice must be used; they appear to be arbitrarily chosen, and do not form a coherent system, but whatever their botanical defects, they do form an agreed and practicable international code.

The following table (Table 1), which is given for reference purposes, is by no means exhaustive, but gives (1) the English names, which are the simplest and clearest for use internally in Britain, (2) the names laid down in the E.E.C. directives, which must be used in all official Community activities, and (3) for comparison, the equally complicated but more logical and coherent system proposed by Mansfeld in 1962,* which is the one used in Field Crop Abstracts. For a fuller discussion of the problem, and references to some of the other proposed systems, the paper by Wellington and Quartley (1972) † should be consulted.

### *Brassica species and animal health*

Glucosinolates (thioglucosides), already referred to as potentially harmful constituents of rape seed (p. 72), may also be present in the vegetative parts of the cruciferous plants, although in smaller quantities than in the seeds. The glucosinolate content of turnips and swedes is usually too low to be a problem, but that of kale may be high enough to produce adverse effects where kale is fed in quantity. Thiocyanates and oxazolidine-thione produced by hydrolysis of glucosinolates are goitrogenic, the former inhibiting the uptake of iodine by the thyroid gland while the latter interferes with the synthesis of the thyroid hormone. Infertility in cattle, which has been associated with kale feeding, may be due to these goitrogenic substances or to the marked mineral imbalance (high calcium–low phosphorus) often found in *Brassica* leaves. Plant breeders are endeavouring to produce kale forms free from these hazards.

Haemolytic anaemia in stock (particularly cattle) may be caused by *s*-methylcysteine sulphoxide present in many of the *Cruciferae,* and found in swede and turnip roots as well as in kale leaves. Changing to a *Brassica*-free diet will usually effect a cure. The high levels of free nitrate ions sometimes found in the leaves of *Brassica* crops grown under a high nitrogen régime may cause nitrate poisoning. Rape

* Mansfeld, R., *Die Kulturpflanze* 2, 1962.

† A practical system for classifying, naming and identifying some cultivated brassicas. Wellington, P. S., and Quartley, C. E., *Journal of the National Institute of Agricultural Botany* **12,** 1972, pp. 413–32.

poisoning, an ill-defined condition of uncertain cause, appears to be associated with the grazing of wet or frosted rape crops.

In spite of these possible hazards, which normally occur only where cruciferous crops form a very high proportion of the diet, the *Brassica* species have been, and no doubt will continue to be, an extremely valuable part of the resources of the livestock farmer.

SINAPIS

The genus *Sinapis* differs from *Brassica* in having the sepals always spreading and the valves of the fruit three- to five-nerved, not one-nerved; the distinction is a small one and the species are often included in *Brassica*; they do not, however, cross with any of the true *Brassica* species.

## *Sinapis alba* L. (*Brassica alba* (L.) Rabenhorst).   **White Mustard**

White mustard is an erect annual, with pinnately-divided, stalked, hairy leaves. The flowers are similar in size to those of the swede; the fruit is a stiffly hairy siliqua with strongly three-nerved valves and a stout, flattened, sword-like beak as long as the valves. The fruits are carried on long, spreading pedicels and stand out almost horizontally. Seed pale yellow, large (2·5 mm). $n=12$. Not native; introduced probably from Mediterranean region.

White mustard (so called from the colour of the seed) is grown for its seed, used in the production of table-mustard in the same way as black and brown mustards, *Brassica nigra* and *B. juncea*. The flavouring compound is, however, different, being pungent tasting but not volatile; it is parahydroxybenzyl isothiocyanate, and is produced by the action of myrosinase on a different glucoside sinalbin, a complex parahydroxybenzyl glucosinolate.

White mustard makes a greater amount of vegetative growth than black mustard, and is more palatable; the seeds do not remain dormant in the soil. It can therefore be used, not only for seed production, but also as a very rapidly-growing catch-crop, either for grazing off in the same way as rape (it must be grazed early, as it rapidly becomes unpalatable when allowed to fruit) or for ploughing in as green manure. The young seedlings are also used in the expanded-cotyledon stage as salad (mustard and cress).*

---

* Cress is the white-flowered annual, *Lepidium sativum*. Rape seed is sometimes substituted for mustard seed.

### Keys for identification of Brassica and Sinapis species

LEAVES:

| | | |
|---|---|---|
| | Leaves glaucous | 1 |
| | Leaves bright or dark green, hairy | 3 |
| 1 | { Leaves all glabrous | *B. oleracea* |
| | { Some leaves with at least a few hairs | 2 |
| 2 | { Young leaves hairy on nerves, upper leaves clasping | *B. napus* |
| | { Leaves slightly hispid, upper leaves stalked | *B. juncea* |
| 3 | { Upper leaves stalked | 4 |
| | { Upper leaves sessile | 5 |
| 4 | { Lower leaves lyrate | *B. nigra* |
| | { Lower leaves pinnate, terminal lobe not large | *S. alba* |
| 5 | { Lower leaves deeply lobed, upper clasping | *B. rapa* |
| | { Lower leaves shallowly lobed, upper not clasping | *S. arvensis* |

FRUIT:

| | | |
|---|---|---|
| | Siliquas smooth, valves one-nerved | 1 |
| | Siliquas rough, valves three-nerved | 3 |
| 1 | { Siliquas long, cylindrical | 2 |
| | { Siliquas short, four-sided, erect | *B. nigra* |
| 2 | { Siliquas spreading | *B. oleracea, napus, rapa* |
| | { Siliquas semi-erect | *B. juncea* |
| 3 | { Beak long, flattened | *S. alba* |
| | { Beak short, conical | *S. arvensis* |

SEED:

| | | |
|---|---|---|
| | Testa pale, yellowish | 1 |
| | Testa dark | 2 |
| 1 | { Seeds large (*c.* 2–3 mm) | *S. alba* |
| | { Seeds small (*c.* 1–2 mm) | *B. juncea* |
| 2 | { Testa brownish-grey | *B. oleracea* |
| | { Testa purple-black | *B. napus* |
| | { Testa red-brown | 3 |
| 3 | { Testa pitted | *B. nigra* |
| | { Testa not pitted | 4 |
| 4 | { Taste mild, some red seeds usually present | *B. rapa* |
| | { Taste acrid, biting | *S. arvensis, B. juncea* |

(Charlock can be distinguished from dark-seeded forms of *B. juncea* by special tests; anthocyanin pigment diffuses out from charlock seed mounted in chloral hydrate.)

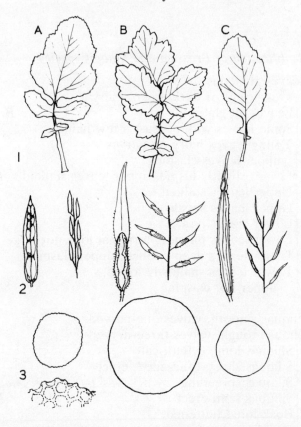

Fig. 16. 1, leaf, × ¼. 2, siliqua, × 1, and diagram of fruiting branch. 3, outline of seed, × 10, of A, black mustard; B, white mustard; C, charlock. A small part of the surface of black mustard seed is also figured, × 50.

### *Sinapis arvensis* L. (*Brassica sinapis* Vis.; *B. sinapistrum* Boiss.). Charlock

Charlock is a weed, differing from white mustard in its rounded, shallowly-lyrate lobed or sinuate leaves, the upper ones sessile; the siliquas with a shorter conical not flattened beak, usually hairy, and less stiffly spreading when mature. Seed small, dark red-brown, closely resembling turnip seed, but with more pungent taste.

### RAPHANUS

The genus *Raphanus* differs from *Brassica* and *Sinapis* in having fruits which, although siliqua-like, are indehiscent. In some species, including *R. rhaphanistrum* L., wild radish, the fruit is constricted between

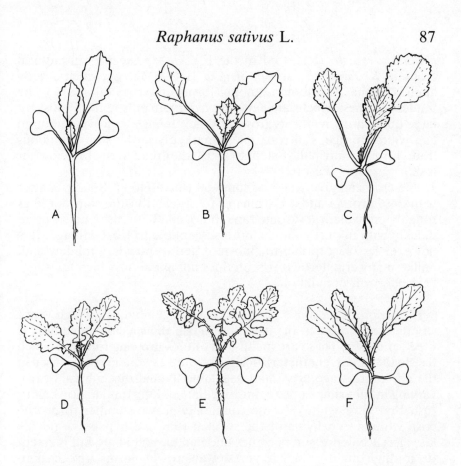

Fig. 17. Diagrams of seedlings of *Cruciferae*, × ½. A, *Brassica oleracea*. B, *B. napus*. C, *B. rapa*. D, *B. nigra*. E, *Sinapis alba*. F, *S. arvensis*.

the seeds and behaves as a lomentum; in the cultivated species the fruit remains whole.

### *Raphanus sativus* L.  **Radish**

An annual or biennial with, in some forms, a bulbous 'root' similar in structure to that of the turnip. Leaves rough, lyrate-pinnate; inflorescence racemose, flowers white or pink to lilac coloured; fruits long, cylindrical, several-seeded, inflated and thick-walled, not breaking up readily when mature except in an irregular manner on threshing; seeds large (*c*. 3 mm), irregular, grey-pink in colour, cotyledons conduplicate. Of uncertain origin; long cultivated, recorded Egypt in third millenium B.C. A wide range of types exists, grown for root, leaf, fruit or seed, but the majority are crops confined to eastern Asia, and two types only are used in Britain.

*Fodder radish.* (Referred in the E.E.C. scheme for agricultural crops to *R. sativus* L. subsp. *oleifera* (DC.) Metzg., which strictly applies to the Asiatic non-bulbing forms grown as oil seeds.) The fodder radishes, which came into use in Britain only in the 1960s, are large quick-growing forms grown for their foliage to provide autumn and winter forage. Cultivars differ markedly in the size of bulbous 'root' formed, and in the extent to which they tend to run to seed after a short growing season.

The crop can give yields of some 60 t/ha at about 8% dry matter with 24% protein and a D-value of 67. It can thus be considered as roughly equivalent to forage rape, but is likely to lose quality more quickly with maturity and is more susceptible to frost damage. It is however less susceptible to club-root and to powdery mildew and, unlike many cruciferous crops, it does not apparently increase sugar beet eelworm populations.

*Horticultural radishes.* (In the E.E.C. vegetable list, which is not strictly comparable to the agricultural list, the specific name only is used for these.) These are mainly very quick growing forms of which the small, tender, characteristically-flavoured swollen 'root' is used in the immature stage as a salad vegetable. Numerous cultivars exist, varying in root shape (globe, intermediate or long) and in skin colour (usually red or white). In the commonly-grown summer forms the crisp white parenchyma of the swollen root and hypocotyl quickly loses its succulence and becomes unpalatable; rapid growth is essential for high quality. Some larger, slower-growing forms are occasionally used as winter radishes; in Asia very large-rooted forms are commonly grown as vegetables.

*Rhaphanobrassica.* Fertile allopolyploid hybrids between *R. sativus* and *Brassica oleracea* have been produced and named *Rhaphanobrassica.* They may perhaps have possibilities as forage crops, and some forms have been in trials in Britain in the 1970s; other hybrids have been considered in Russia as possible vegetables for human consumption.

### OTHER CRUCIFEROUS CROP-PLANTS

*Crambe maritima* L., seakale, is a perennial, the blanched shoots of which are used as a vegetable. Fruit indehiscent, one-seeded, cotyledons conduplicate.

*Lepidium sativum* L., cress, is a white-flowered annual, with fruit a

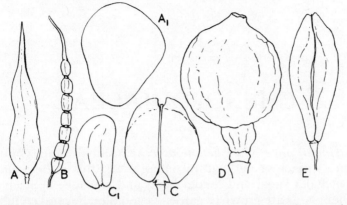

Fig. 18. Seeds and fruits of various *Cruciferae*. A, pod, × 1, and A$_1$, seed, × 7, of cultivated radish. B, wild radish pod, × 1. C, silicula, × 4, and C$_1$, seed, × 7, of cress. D, indehiscent fruit of seakale, × 4. E, indehiscent fruit of woad, × 2.

two-seeded silicula; cotyledons three-lobed, radicle incumbent. Used mainly in seedling stage as 'mustard and cress'. *Isatis tinctoria* L., woad, a biennial with large panicles of small yellow flowers; was formerly much cultivated for dye-production. Fruits one-seeded, winged, indehiscent; radicle incumbent.

*Armoracia rusticana* Gaertn., horse-radish, is a perennial grown for its strongly-flavoured tap-roots. Flowers white, fruit a few-seeded silicula, radicle accumbent; seed not ripened in Britain. *Nasturtium officinale* R. Br., water-cress, is an aquatic perennial with white flowers and a slender siliqua with seeds in two rows in each cell; diploid. The brown winter variety is a sterile triploid derived from a cross between this and the tetraploid *N. microphyllum* (Boenn.) Rchb., with seeds in a single row.

# 4

# CHENOPODIACEAE

*General importance*. The *Chenopodiaceae* is a small family containing only one species of agricultural importance; this is *Beta vulgaris,* which includes sugar and fodder beets and mangels. Spinach and a few other plants are grown as garden vegetables. Several members of the family are common arable weeds; these include species of *Chenopodium,* from which the family name is derived.

*Botanical characters*. Herbs and small shrubs, often halophytic. Leaves simple, often with 'mealy' covering of short, swollen hairs, alternate, exstipulate. Flowers small, inconspicuous, wind-pollinated, often in dense clusters. Perianth usually of five segments, stamens equal in number or fewer, placed opposite the perianth segments. Ovary of two or three united carpels, superior except in *Beta,* unilocular with a single basal curved ovule. Seed with embryo curved around the starchy food reserve tissue, which is mainly perisperm (the remains of the nucellus), not true endosperm.

### *Beta vulgaris* L.   **Beet**

The wild forms of beet are sea-coast plants of Europe and Asia, very variable in habit and duration and perhaps referable to a number of different species. The cultivated forms, probably derived from eastern Mediterranean types, are biennials, grown either for their fleshy leaves, or more commonly for the swollen 'root'.

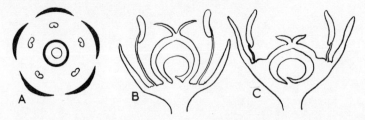

Fig. 19. *Chenopodiaceae.* A, floral diagram. B, diagram of vertical section of typical flower. C, of *Beta.*

*Life-history*

*Seedling.* Germination is epigeal; the two cotyledons are almost sessile and lanceolate, with blunt tips. The hypocotyl is elongated and distinctly stouter than the root; the epicotyl remains short and a rosette of glabrous dark-green leaves develops. These leaves are ovate, tapering to a long, rather broad petiole.

*First-year plant.* The primary root is diarch, and two vertical lines of lateral roots are produced. Secondary thickening commences in the

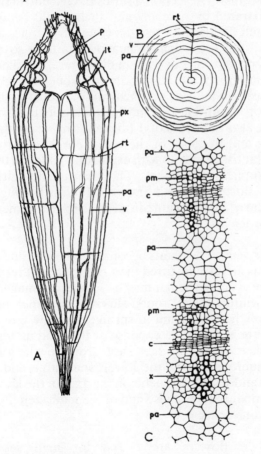

Fig. 20. A, diagram of longitudinal section of 'root' of a beet (long shape). B, transverse section. C, part of transverse section at higher magnification, to show arrangement of cells. *c*, cambium. *lt*, leaf-trace. *p*, pith. *pa*, parenchyma. *pm*, phloem. *px*, primary xylem, closely surrounded by secondary xylem formed by first cambium. *rt*, root-trace. *v*, vascular ring. *x*, xylem.

usual way, by the development of a cambium between the primary xylem and primary phloem, but after this it proceeds in an unusual manner. Instead of the one cambium continuing to grow and produce more and more secondary tissue, a second cambium arises in the pericycle. The xylem produced by this cambium forms a ring outside the phloem from the first cambium, and separated from it by a ring of parenchyma. A third cambium arises outside the second, and then a fourth, until some eight or nine cambial rings have developed, each producing xylem internally and phloem externally. The mature root thus shows in transverse section a series of concentric rings of vascular tissue, separated by parenchyma. In the wild beets, and in the forms cultivated for their leaves, the concentric zones are comparatively narrow, and the xylem is heavily lignified, so that a rather woody root is produced, 2 or 3 cm in diameter. In mangels and other cultivated forms grown for their 'roots', the zones are very broad, and the xylem consists mainly of parenchymatous cells, so that a massive succulent 'root' is produced. The thickening involves not only the true root, but also the hypocotyl (distinguishable by the absence of lateral roots), and the base of the true stem, where the concentric rings of vascular tissue link up with each other and with the leaf-traces to form a complicated network. The stem in the cultivated forms remains very short during the first year (unless bolting occurs) and forms the 'crown' of the plant; from it arise the numerous closely-crowded large leaves.

*Second-year plant.* The plants become dormant in late autumn, and the 'root' forms are usually lifted then. They are not resistant to hard frost, and mangels in particular may be severely damaged if left in the ground over winter. If the 'roots' survive, or if they are lifted and replanted, growth starts again in spring; this new growth is largely elongation of the stem at the expense of the food material stored in the previous year. A stout, ridged stem grows to a height of 1·5–2 m and bears numerous leaves, the lower ones large and stalked, the upper smaller and sessile. Branches in the axils of the leaves also grow out, and the upper parts of these and of the main stem form long, lax, spike-like inflorescences.

*Flowers.* The flowers are borne in small sessile clusters (glomerules) in the axils of bracts. Each flower has a perianth of five greenish-yellow segments, and five stamens arranged opposite the perianth segments. The ovary is almost completely inferior and bears usually three short styles. Pollination is largely by wind, and cross-pollination is usual. The perianth does not wither and drop off as the

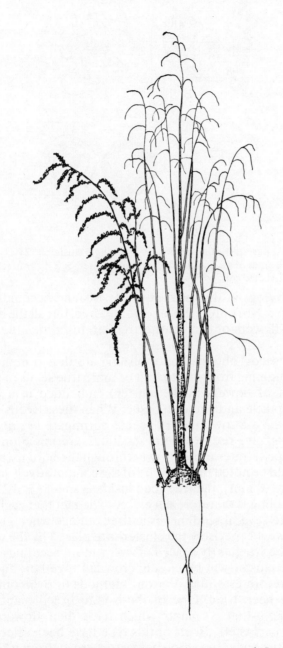

Fig. 21. Second-year mangel plant, in fruiting stage, × $\frac{1}{11}$. Leaves omitted; fruit-clusters shown on some branches only.

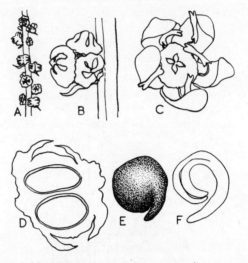

Fig. 22. Beet. A, part of fruiting branch, × ½. B, single cluster, × 2½. C, individual flower, × 7. D, diagram of section through mature cluster, × 5. E, true seed, entire, and F, embryo dissected out, × 7.

single seed develops in the ovary, but persists and becomes almost woody in texture. Not only is the seed not set free, but all the enlarged and hardened flowers of each cluster remain fused together.

*'Seed'*. The 'seeds' obtained on threshing are thus complex structures, each containing from one to four or more true seeds (according to the number of flowers in the cluster) embedded in a mass of hardened receptacle and perianth tissue. When these 'seeds' or clusters (glomerules) are sown the true seeds germinate *in situ*, so that each cluster may give rise to several seedlings. Germination may be slow, as water penetrates rather slowly through the hard tissue of the cluster. It should be noted that, in spite of the comparatively large size of the cluster (3–8 mm), the true seed inside is small (1·5–2·5 mm), and shallow drilling is therefore necessary. The fact that each cluster may give rise to several seedlings is a disadvantage when growing a crop, since however precisely the clusters are placed by the drill, the seedlings will be irregularly spaced. Two or more seedlings derived from the same cluster will be closely crowded together, and hand-singling is therefore essential. Various methods of overcoming this difficulty have been tried. One of these is to breed plants giving *'monogerm seed'*—that is, plants which have their flowers borne singly and not in clusters. Beets of this type have been selected, the bulk of the British sugar beet crop being now grown from monogerm 'seed'. The rather irregular lens shape of the 'seed' makes it unsuit-

able for precision drilling and monogerm sugar beet 'seed' is always pelleted and graded before sale. Monogerm cultivars of other forms of the beet species are becoming more readily available. The alternative is some form of mechanical treatment of the usual multigerm clusters. No method of releasing the true seeds from the clusters has been devised, and the treatments adopted have been segmenting and rubbing. *'Segmented seed'* is produced by chopping the clusters into smaller pieces, *'rubbed seed'* by rubbing them against an abrasive wheel. In both cases the treated 'seed' is then graded by sieving so as to retain only those fragments likely to contain a single true seed. The rubbing method is the more popular in Britain. The rubbed 'seed' is usually graded from 2·8–4·4 mm and pelleting, to ensure suitability for use in precision drills, is normal.

Germination of beet clusters is expressed as a percentage of the clusters (or piece of cluster, in the case of rubbed seed) which produce at least one seedling. Additional seedlings from the same cluster are of no value, and for this purpose can be ignored. Germination of clusters is rarely above 85%. Untreated 'seed' is extremely variable in weight, but has usually 33 000–44 000 clusters per kilogram; rubbed and graded (unpelleted) 'seed' about 65 000–70 000.

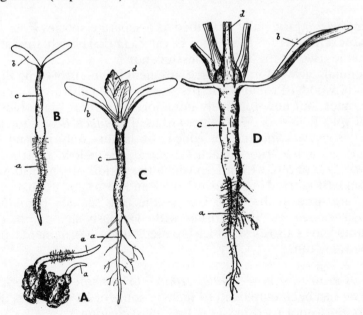

Fig. 23. Mangel. A, germinating cluster; B, C, D, stages in the development of a single seedling. *a,* root; *b,* cotyledon; *c,* hypocotyl; *d,* first foliage leaves. From Percival, *Agricultural Botany.*

*Range of types*

The forms within the species can be divided into the wild sea beet (subsp. *maritima* (L.) Thell., originally described as a separate species, *B. maritima* L.), and the cultivated forms (subsp. *vulgaris*), which are again divisible into the forms grown for their leaves and those grown for their roots.

*Wild sea beet.* Usually perennial, low-growing, with branching prostrate stems arising from a woody root, and rather small angular leaves. Inflorescences like those of cultivated forms, but usually shorter, more or less prostrate, and with smaller flower clusters. A common seashore plant in Britain, extending in various forms to the shores of most of Europe and West Asia. Annual wild forms exist in coastal regions of southern Europe and may have given rise to the weed beet (annual beet, wild beet) now causing problems in sugar beet crops in Europe and parts of Britain. Some authorities consider, however, that these weed beets have evolved directly from bolters in sugar beet crops and not by invasion of, or by cross-pollination with, existing wild forms.

*Leaf beets.* (Originally described as a separate species *B. cicla* L., and now referred to as *B. vulgaris* L. var. *cicla* (L.) Ulrich; this latter is the name used in the E.E.C. classification.)
Biennials, grown for their large succulent leaves; roots only slightly swollen, woody. Probably the first forms to be cultivated in prehistoric times, but now relatively unimportant garden vegetables, not grown on a field scale. Two types are used: *Spinach Beet*, sometimes called *Perpetual Spinach,* in which the lamina only is used, as a substitute for the true spinach (*Spinacia*, see below, p. 103), and *Seakale Beet* or *Swiss Chard,* in which the petiole and midrib, which are white (bright red in one form), thick and fleshy and up to 8 or 9 cm wide, are used in the same way as the true seakale (*Crambe,* in *Cruciferae*, see p. 88). Forms with variously-shaped, brightly-coloured leaves also exist, which are used purely as ornamental plants for bedding-out.

*Beets grown for their swollen 'roots'.* (The name *B. vulgaris* var. *rapacea* has been employed to include some or all members of this group, but is not used in the E.E.C. classification.)
A large and variable group of biennials of considerable importance, in which specialized forms have been selected for use for three distinct purposes: (1) for human consumption—beetroot; (2) for the

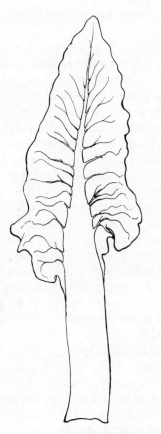

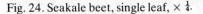

Fig. 24. Seakale beet, single leaf, × ¼.

extraction of sugar—sugar beet; (3) for stock-feeding—mangels and fodder-beet.

(1) **Beetroot** (*B. vulgaris* var. *esculenta* L. in the E.E.C. classification) is the name used for the forms grown as root-vegetables for human consumption. They produce succulent 'roots' with very little hard lignified tissue. Recorded from the Roman period; in the past various different colour forms have been used, but at present (apart from one or two white and orange novelty forms) only deep red forms are grown. In these, the cell sap contains anthocyanin pigments, and selection is aimed at producing 'roots' of even, very dark red colour, with as little difference as possible between the rings of vascular tissue and the intervening parenchyma, both in colour and texture. In many varieties the red coloration extends to the leaves and also to the second-year stem. Half or more of the swollen 'root' usually stands above ground-level; long and intermediate varieties are grown for

main crops for winter use, globe and flat varieties for earlier production and forcing.

(2) **Sugar beet** (the name *B. vulgaris* var. *saccharifera* has been used for this form, but is not employed in the E.E.C. scheme, which lists all the agricultural forms of types (2) and (3) merely under the specific name *B. vulgaris*) is a specialized type grown for processing in factories for the extraction of sugar. It was developed at the end of the eighteenth century in Europe from the white Silesian beet, which was then grown for fodder, and which was found to be the most suitable source of sugar as an alternative to the tropical sugar-cane (*Saccharum,* in the *Gramineae*). The strategic blockade of the Continental ports during the Napoleonic Wars cut off the supply of cane sugar from the West Indies, and thus favoured the development of an alternative source of sugar. The original forms contained only about 4% of sugar; careful selection, followed more recently by intensive breeding work, has resulted in the raising of this figure to a maximum of approximately 20%. This, combined with a steady improvement in methods of sugar extraction, has resulted in sugar beet becoming one of the outstanding arable crops of most temperate regions. In Britain, where it is primarily suited mainly to the warmer areas of the south and east, its cultivation dates from 1912.

Sugar beet plants have white 'roots' of conical shape, growing deep in the soil with only the crown exposed. They usually show two shallow, vertical grooves (the so-called 'sugar-grooves') in which the two lines of lateral roots emerge. They tend often to be somewhat irregular in shape. Any pronounced irregularity or branching of the root is undesirable, as it results in greater adherence of soil to the root when lifted, and may make processing less efficient. The 'roots' are small compared with mangels, usually weighing from 0·5–1 kg. Larger roots tend to have a lower sugar content. The highest sugar concentration is associated with the phloem of the vascular rings and, other things being equal, 'roots' with numerous narrow rings show the highest sugar content.

The lower percentage sugar content of large roots means that the number of plants per hectare has an important effect on the total yield of sugar. Plant populations of over 90 000 per hectare have given the highest yields of sugar in trials but such close spacings are difficult to achieve in commercial practice and populations from 73 000 to 86 000 per hectare are aimed at. Sugar beet is commonly drilled in rows about 500 mm apart using seed-rates which vary considerably depending on type of 'seed' (pelleted or unpelleted, monogerm or multigerm), whether drilled to a stand or to be singled, type of drill used and size grade of the 'seed'. The rate may vary from 8–30 kg/ha for pelleted 'seed' and from 1–12 for unpelleted. It is essential that a good seed-bed should be provided otherwise germination and establishment may be low and irregular, resulting in low plant populations. If the plant population falls much below 70 000 per hectare the increase in the size of individual roots fails to compensate for the reduced plant numbers and both the root yield and the sugar yield per hectare will be reduced.

The time of sowing also has a marked effect on yield. In general, the earlier the 'seed' is sown the higher is the yield, providing that 'bolting' does not take place. The flowering of beet is influenced both by temperature and day length (cf. vernalization of cereals). Plants change from the vegetative to the flowering condition when the day becomes sufficiently long (the necessary length varies in different strains), providing that the plant has passed through a preliminary period of exposure to low temperatures. With normal sowings this cold period is not experienced until the following winter, and the plant flowers during the second year. If, however, seed is sown very early, and cold weather follows sowing, the plant behaves as an annual, flowering during the first year—that is, 'bolting'. The useful yield of plants which bolt early in the season may be only half that of normal plants, and a high percentage of bolters therefore materially reduces the sugar yield per hectare. Sowing before mid-March is likely to give, in most years, a high proportion of bolters, while late-sown crops are usually equally unsatisfactory since, although no bolting takes place, the yield is reduced by the shorter growing season.

*Sugar beet cultivars.* Many cultivars of sugar beet exist, all rather similar in appearance and differing mainly in physiological characters. The comparatively recent introduction of sugar beet growing into Britain has meant that Continental varieties, largely of German origin, have been commonly grown; more recently varieties bred in this country have come into widespread use. The current N.I.A.B. system is to group varieties into monogerm and multigerm varieties without further subdivision. Most cultivars are diploid but several polyploid (triploid) varieties have been developed. In sugar beet triploids tend to be more vigorous than either diploids or tetraploids but those available do not necessarily surpass the best diploid varieties. Triploids are largely sterile and fresh triploid seed has to be produced each year by interplanting rows of male-sterile seed-producing diploid plants with rows of fully fertile tetraploids, seed being harvested from the diploid plants only.

Almost all varieties now used are capable of giving root yields of some 40 tonnes per hectare giving some 6–7 tonnes of sugar per hectare; the sugar content varies from 15·5 to 18% depending on variety, season and soil conditions.

Varieties show marked differences in resistance to bolting. The variety Vytomo, the best variety from this point of view in the N.I.A.B. Recommended List for 1977/8, gave 0·1% of bolters from a normal sowing date and 1·1% from an early sowing, whereas for the variety Sharpes Klein Monobeet the corresponding figures were 0·7% and 4·9%. Varieties also differ in size of tops, i.e. of the crown and attached leaves which are cut off at harvest. Yield of tops varies from about 30 to 40 t/ha. These tops, which have a nutrient value approaching that of kale, may be fed to stock after wilting, or as silage, and are a valuable by-product of the sugar beet crop.

The yield of dry matter per plant is not so much greater in the large-topped forms as might be expected, since the small-topped forms have a higher net assimilation rate (amount of dry matter produced per unit leaf area); this partially compensates for the difference in leaf area. Cultivars also differ in their genetic resistance to downy mildew and in their tolerance of infection by virus yellows.

(3) **Mangels*** and **fodder beet.** These are the various forms of beet with swollen 'root' grown for stock-feeding. They might thus logically all be described as fodder beet, but the very distinct large-rooted forms known as mangels have a much longer history in Britain than the others; they were apparently developed in central Europe by the sixteenth century, and were introduced into Britain in the eighteenth. They have thus retained their name, while the term fodder beet is used in Britain only for the more recently developed smaller-rooted types intermediate in character between mangels and sugar beet.

*Mangels.* The typical mangel is a beet with a very large 'root', of which a high proportion is derived from the hypocotyl, and which stands high out of the ground, with only about one-third of the 'root' below ground level. It is therefore easily lifted, but is more readily damaged by frost than sugar beet, and must be lifted and clamped early.

In Britain the breeding of mangels has in the past been directed mainly towards the production of large 'roots' (averaging perhaps 2–3 kg each), easily lifted and of regular shape, with small tops. The 'root' shape is usually globe or tankard, less commonly long or intermediate; the skin colour is red, orange or yellow, while the 'flesh'

*Also written 'mangolds'. *Mangold* is a German name of uncertain derivation for leaf-beet or chard; the forms with swollen 'root' appear to have been named from this *Mangold-wurzel* (i.e. chard root), and this later misinterpreted as *Mangel-wurzel* (i.e. scarcity root). The spelling 'mangel' is to be preferred, in spite of its derivation, since it avoids the possibility of confusion with the German name for leaf-beet.

often shows concentric rings of colour, the vascular rings being white and the intervening parenchyma of the same colour as the skin. Agricultural varieties are usually named according to their shape and colour, e.g. Long Red, Golden Tankard, etc.; most varieties have a dry matter content of 10–12%.

In Denmark and some other Continental countries mangels have been developed along rather different lines; selection has been based largely on dry-matter content of the 'roots', rather than on size and ease of lifting. The resulting medium dry matter mangels, with an average dry matter content of 12–15%, are mainly of intermediate shape, with about half the 'root' above ground. The size of 'root' and the root yield per acre is smaller than with English mangels, but this is more than compensated for by the lower percentage of water in the 'roots'; the yield of dry matter per hectare is therefore greater. The tops, which in Denmark are used for fodder, are larger than those of the English mangels, in which the leaves have usually been discarded.

Both types of mangels are grown primarily for storing and feeding to cattle in late winter and spring; they cannot be folded off, as the 'roots' are likely to cause digestive disturbances in autumn, and they can only be safely fed after a period of 'ripening' in store.

*Fodder beet.* A considerable number of types of beet exist inter-mediate in characters between mangels and sugar-beet. These were developed particularly in Denmark and Holland and have been used in those countries since about 1930; they were not introduced into Britain until after 1940. They show a continuous range of variation from forms which are essentially special selections of sugar beet grown for stock-feeding to ones which differ little in appearance from medium dry matter mangels. All have large tops, and the 'roots' are

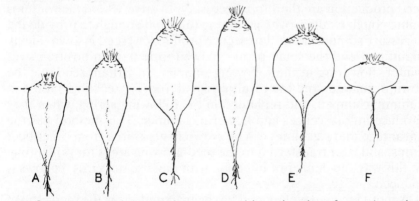

Fig. 25. Diagrams of shapes of mature 'root' in various beet forms. A, conical (sugar beet). B, intermediate (most fodder-beets). C, tankard. D, long. E, globe (mangels and some red beet). F, flat (red beet only).

suitable for feeding from autumn onwards, without the period of after-ripening necessary for mangels.

They have dry matter contents varying from about 15 to 20%; those with the lower figures (medium dry matter) being suitable for cattle and those nearer to sugar beet (high dry matter), with firmer roots, more suitable for pigs. The fodder beets give higher dry matter yields than mangels but are generally more difficult to lift as the roots are more deeply buried. They are important mainly as alternatives to cereals and were popular in the immediate post-war years; they went out of common use as grain became more freely available.

Monogerm cultivars have been produced, and these coupled with the likely development of efficient lifting and handling machinery might well lead to a revival of interest in this crop. Dry matter yields of 10 t/ha can be obtained from the roots, with another 5 tonnes from the tops. This compares well with mangels which give a root yield of some 8–8·5 tonnes of dry matter per hectare, and usually have smaller tops than fodder beet.

### Seed production in beets

All forms of the species *Beta vulgaris* are mainly cross-pollinated, and will all intercross freely. It is therefore essential that crops of different forms being grown for seed should be well separated. In order to ensure this, a system of zoning has been developed in the seed-growing areas.

The cultivated forms, being biennials, usually flower in the second year, but in most cases plants grown in the normal way cannot be safely over-wintered in the field. Special methods of growing beet for seed production are therefore necessary. In areas where the winter is mild enough to allow over-wintering (but cold enough to provide the necessary stimulus for inflorescence production) seed is sown thickly in autumn and the small plants allowed to remain in position over winter, flowering in the following summer. In Britain seed may be sown in summer and the resulting small 'roots' (stecklings) lifted in autumn, clamped, and replanted in the following spring, when they produce inflorescences and flower in summer. This method has the advantage that stecklings can be grown in areas free from other beet crops, and later transferred to the seed-growing areas for replanting. In this way the danger of infection with the disease Virus Yellows is reduced.

A more recent development has been to undersow the seed crop in barley, which acts as a barrier against infection in the first year; spraying with systemic insecticides to control the aphid vectors of

Virus Yellows may also be used. Much of the seed crop now grown is produced in this way. The higher plant population obtained by under-sowing in barley, as compared with transplanted stecklings, gives a greater seed yield per hectare and the smaller size of individual plants so obtained makes for easier seed harvesting. Where the initial mother seed stock is very limited the steckling method may still be used as this gives larger flowering plants and hence larger seed yields per plant.

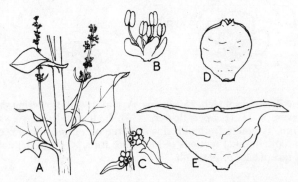

Fig. 26. Spinach. A, part of male plant in flowering stage, × ½· B, single male flower, × 4. C, cluster of female flowers, × ½. D, E, mature 'seed' of round and prickly seeded forms, both × 4.

### *Spinacia oleracea* L.    **Spinach**

An annual, grown as a vegetable for its thick, succulent leaves. The broad, dark-green, often crinkled leaves form a rosette in the young plant; later an erect leafy stem is produced, up to 0·6 m high. The flowers are borne in clusters in the axils of the upper leaves; they differ from beet in being unisexual, the male flowers with four to five perianth segments, and four to five stamens, the female with a two- to four-toothed perianth surrounding a single, one-seeded ovary with four to five short styles. Spinach is usually dioecious, but occasional monoecious plants occur. The perianth of the female flower hardens and persists around the indehiscent fruit, and the 'seed' thus has approximately the same structure as a monogerm beet 'seed'. The 'seed' is smooth in *round* or *summer spinach*, or rough owing to the projecting calyx-teeth in *prickly* or *winter spinach*. The latter is more hardy and can be used for autumn sowings to stand over winter, but is more liable to 'bolt' in hot weather. Even in summer spinach the period during which the plant remains in the usable vegetative con-dition is short, and frequent successional sowings are necessary. Perhaps of Persian origin; not recorded before about A.D. 500.

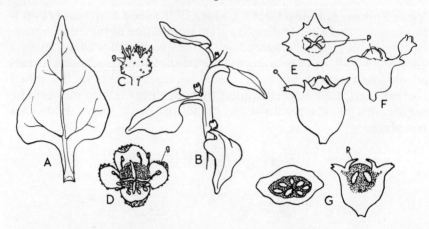

Fig. 27. New Zealand spinach, *Tetragonia*. A, leaf, × ½. B, flowering shoot, × ½. C, immature flower, × 1½. D, open flower, × 3½. E, top and side views of mature indehiscent fruit. F, young fruit with flower borne on outgrowth of receptacle. G, transverse and longitudinal sections of immature fruit, × 1½. *g*, glandular hairs. *o*, outgrowth of receptacle, *p*, perianth.

*Chenopodium bonus-henricus* L., Mercury or Good King Henry, a low-growing perennial, and *Atriplex hortensis* L., Garden Orache, a tall annual with pale green (or in one variety, deep-red) leaves, are two other members of the family occasionally grown, and used in the same way as spinach. *Chenopodium quinoa* Willd. is an annual grown in S. America for its abundant seeds, used in the same way as cereals.

*Tetragonia tetragonoides* (Pallas) O. Kuntze (*T. expansa* Murr.), New Zealand spinach, belongs to a distinct but related family, the *Tetragoniaceae,* formerly included in the *Aizoaceae.* It is an annual with thick, fleshy, triangular, alternate leaves, and small greenish-yellow axillary flowers, with four-partite perianth, numerous stamens and semi-inferior ovary. Several ovules are present; the receptacle develops pointed, horn-like projections and hardens around the mature ovary to give a large (1 cm) almost woody, irregularly-shaped 'seed', which absorbs water very slowly. Discovered in New Zealand in 1770, coming into use in Britain about 1820, and thus one of the very few comparatively recently introduced species used as a vegetable. More tolerant of heat and drought than spinach, and occasionally replacing it for summer use.

# 5

# UMBELLIFERAE

*General importance.* The *Umbelliferae* are not of great agricultural importance, but include a number of plants grown as vegetables for human consumption and occasionally as root-crops for stock-feeding. Many members of the family contain strongly-scented resin-like substances, and are used as flavouring herbs. A few contain alkaloids and are important poisonous weeds.

*Botanical characters.* A large family with between 2 000 and 3 000 species, almost all very uniform in structure, so that the family (like the *Cruciferae*) is usually easy to recognize, but distinction within the family is often difficult. Nearly all are herbaceous plants with erect, hollow stems and alternate exstipulate leaves, often much divided and with broad sheathing bases to the petioles. The flowers are small, but massed together in conspicuous umbels. The umbels (from which the family takes its name) are usually compound; a whorl of bracts (*involucre*) is sometimes present at the base of the primary umbel, and whorls of bracteoles (*involucels,* or partial involucres) at the base of the secondary umbels. The individual flowers consist of five sepals, which are often reduced to minute points or may even be completely suppressed, five free petals, five stamens and two joined inferior carpels. A single pendulous anatropous ovule is present in each of the two chambers of the ovary. The ovary is surmounted by a fleshy disk (*stylopodium*), around which nectar is secreted; the two short styles are borne on the disk. Pollination is usually by small flies; selfing is largely prevented by marked protandry. The fruit is a schizocarp, splitting when ripe into two single-seeded mericarps which may remain for a time attached to a slender branched 'stalk' (*carpophore*). It is these mericarps or half-fruits which form the 'agricultural seed' in the *Umbelliferae*. They vary considerably in shape, and classification of genera and species within the family is very largely based on them. The inner (*commisural*) surface by which the two mericarps were attached is usually somewhat flattened, the outer surface usually convex with five longitudinal ridges, between which lie oil-canals (*vittae*). Within the rather thick, leathery wall of the mericarp is the

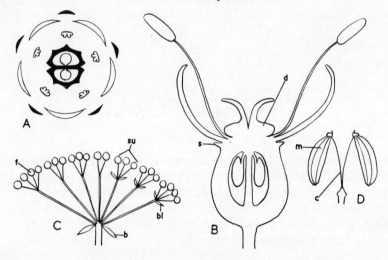

Fig. 28. Floral characters of *Umbelliferae*. A, floral diagram. B, diagram of vertical section of flower. C, diagram of compound umbel. D, fruit separating into two mericarps. *b,* bract of involucre. *bl,* bracteole of involucel. *c,* carpophore. *d,* disk (stylopodium). *f,* individual flower. *m,* mericarp. *s,* sepal (often much reduced or absent). *su,* single secondary umbel.

single seed, with abundant endosperm in which is embedded the small, straight embryo.

### *Daucus carota* L.   **Carrot**

A biennial with tap-root which is woody in the wild form, thick and succulent in the cultivated forms. Leaves thrice-pinnate with small lanceolate pointed lobes, forming a rosette in the first year. Stem lengthening in the second year, ridged, solid, leafy, up to 1 m high, bearing numerous compound umbels which are flat or slightly convex in flower, and deeply concave in fruit, giving a 'bird's nest' appearance. Bracts usually from three to six, divided into several narrow, pointed lobes. Individual flowers small, petals notched, white (petals of central flower of umbel sometimes reddish). Fruit ovoid, primary ridges small, bearing short, stiff hairs. Secondary ridges are present between the primary ridges, and are much more conspicuous, forming on each mericarp four lines of stiff, hooked spines.

#### *Life-history*

'*Seed*'. The spiny mericarps which are the structures obtained on threshing a carrot-seed crop are difficult to deal with, as they mat

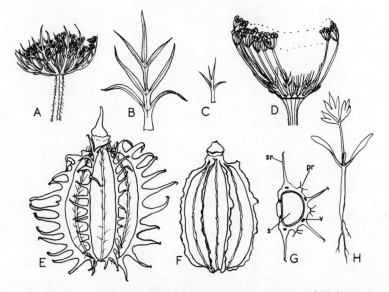

Fig. 29. Carrot. A, small compound umbel in flowering stage, × ½. B, bract, × 1. C, bracteole, × 1. D, umbel in fruiting stage (part omitted), × ½. E, mericarp in unrubbed state, × 10. F, after rubbing. G, diagrammatic transverse section of E. H, seedling, × 1. *pr*, primary ridge. *sr*, secondary ridge. *s*, seed. *v*, vitta.

together and cannot be drilled readily. They are therefore 'rubbed' during the cleaning process; this does not set free the true seed, but rubs off the spines and part of the ridges, and the 'seed' normally sown consists of the rubbed and more or less smoothed, roughly hemispherical mericarps, about 3·5 mm long, about 1–1·5 million/kg; pelleted seed is used for precision drilling. Germination is usually low, and 70% is accepted as satisfactory.

*Seedling.* Germination is slow. The cotyledons are elongated, strap-shaped, the first foliage leaf deeply divided and about as long as broad. The succeeding leaves are much longer. The hypocotyl and the upper part of the root become swollen; the epicotyl remains very short.

*Structure of swollen root.* The primary root is diarch; four vertical rows of lateral roots are produced. These are not confined to the true root, but adventitious roots are produced on the hypocotyl, which in most varieties remains rather short and is pulled down into the ground. Normal secondary thickening takes place, but the secondary xylem is largely unlignified, and the secondary phloem forms a wide zone around it. The structure is thus similar to that of a turnip, but the

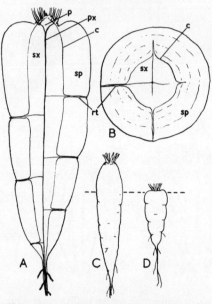

Fig. 30. Carrot. A, diagrammatic longitudinal section of root at end of first year's growth. B, transverse section. C and D, comparison of habit of cattle carrot (White Belgian) and garden carrot (Intermediate). *c*, cambium. *p*, pith. *px*, primary xylem. *rt*, root-trace. *sp*, secondary phloem, *sx*, secondary xylem.

proportion of phloem to xylem is very much greater, and only root and hypocotyl are involved.

*Second-year growth.* Occasional plants may 'bolt', producing inflorescences in the first year and then dying, but the majority of plants behave as biennials. They are usually lifted and stored at the beginning of the winter dormant period, as they are easily damaged by frost. Growth recommences in the spring, new leaves are produced, and further secondary thickening takes place. The secondary xylem produced in this second year is heavily lignified, and the feeding value of the root therefore decreases when new growth starts. If the root has been left in the ground the stem (epicotyl) elongates, and the reserve food stored in the root is used in the production of the inflorescence and fruit. The whole plant dies when the fruit is ripe.

### Range of types

The wild form *D. carota* subsp. *carota* (wild carrot) has a rather slender white or purplish woody tap-root, and occurs as an occasional weed in Britain. The cultivated carrots, subsp. *sativus* (Hoffm.)

Hayek, have been in cultivation for over 2 000 years, and are perhaps derived from western Asiatic forms of the species. They vary in colour, size and shape of root. The original cultivated carrots were apparently purple rooted owing to the presence of anthocyanin pigments, but white and yellow anthocyanin-free mutants were selected and came to be preferred. The orange or red carrots now grown are derived from forms selected for a more orange colour in Holland during the seventeenth and eighteenth centuries. Some large-rooted white and yellow forms remained in use as cattle carrots for stock-feeding. The varieties grown for human consumption are all red carrots, which have about 13% dry matter, of which some 40% consists of sugars. Carotene, which is largely responsible for the colour and is nutritionally valuable as a precursor of vitamin A, is present to the extent of about 0·25% of the dry weight. Selection in carrots for human consumption is directed towards the production of roots with as little 'core' as possible —that is, with the secondary xylem as similar as possible to the phloem both in texture and colour. Fungal diseases are not important; the main pest is the carrot-fly, *Psila rosae,* but no varieties are known which show resistance to the attacks of this insect. Motley dwarf virus may cause serious loss.

### Cultivars

*Red carrots.* Cultivars are classified into groups on time of maturity and on shape of root, which may be very short and almost globular, cylindrical, tapering, long or short, with the tip blunt (stump-rooted) or pointed. Cylindrical varieties are preferred for pre-packing and varieties in which the xylem core is as deeply pigmented as the outer phloem zone (red-core types) are considered superior to yellow-cored varieties. Well-known groups of cultivars are:

Amsterdam Forcing: early maturing with cylindrical roots, grown for marketing bunched or pre-packed.

Nantes: somewhat larger cylindrical roots grown for pre-packing, canning, or fresh sale.

Chantenay: conical roots with good red-core character; small roots canned whole, larger ones sliced or diced (canned or dehydrated) or sold fresh.

Berlicum: later maturing good quality cylindrical roots grown for pre-packing and market.

Autumn King: late maincrop, high yielding with large tapering roots, colour often inferior to other groups; keeps well and used mainly for fresh sales throughout the winter.

Carrots for canning whole or for pre-packing are commonly grown in narrow rows (*c.* 9 cm) with seed-rate and sowing dates adjusted to give a high proportion of the crop in the size range required. Total yields of marketable roots may be some 25 tonnes per hectare for earlies and 50 tonnes for maincrop varieties.

*Cattle carrots*. White Belgian has large tapering roots with white flesh, standing out of the ground to a height of 10 cm or more, the exposed part often greenish. This form can give the highest total fresh weight yield, although of rather lower feeding value, and commonly gave over 50 t/ha at a period when this was a very high yield; it was formerly used as a field crop for stock food, but has now largely gone out of use. Yellow Belgian was similar but with yellow flesh and somewhat higher feeding value. Carrots are now very rarely grown in Britain for stock food, although surplus or unsaleable roots may be used in this way.

## *Pastinaca sativa* L. (*Peucedanum sativum* (L.) Benth.).   **Parsnip**

Parsnips differ from carrots in their larger, simply-pinnate leaves with ovate segments, taller, hollow flowering stems and umbels with bracts and bracteoles few or absent, with the rays of the umbel not incurved in fruit. The petals are yellow and the fruit larger (6–7 mm long), much flattened and without secondary ridges. The mericarps are thus thin, flat and oval, 150 000–250 000 per kg. They do not require milling before sowing, but the flat shape is unsuited to drilling and 'seed' may be treated with French chalk for easier flow or pelleted for precision drilling. The germination is always low, and parsnip 'seed' cannot be safely kept for more than one year. The seedling is similar to that of the carrot, but larger, with broader cotyledons and less-divided foliage leaves.

Wild parsnip (subsp. *sylvestris* (Mill.) Rouy and Camus) is an unimportant biennial grassland plant of some limestone soils in Britain; it has a slender, rather woody tap-root. The cultivated forms (subsp. *sativa*) have swollen roots similar in structure to those of carrots, but with an even larger proportion of secondary phloem. Only white (really pale yellow) forms exist. Parsnips were formerly grown to some extent for stock food on soils heavier than those suited to carrots, and gave yields of about 50 t/ha. The dry matter is about 15%, with a higher starch content than carrots; prior to the development of fodder beets parsnips had the highest nutritive value of all fodder root-crops. Their use is now confined to growing as a winter vegetable for human consumption. Only long (e.g. Hollow Crown)

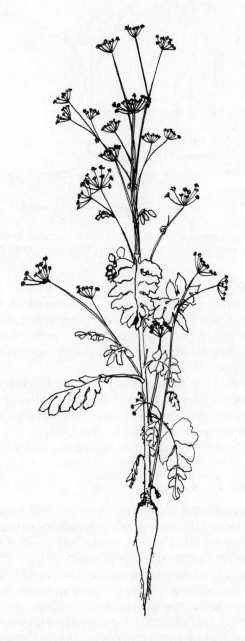

Fig. 31. Parsnip. Small second-year plant in unripe fruit stage, × 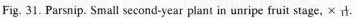.

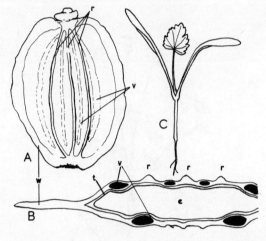

Fig. 32. Parsnip. A, mericarp, × 7. B, part of transverse section of mericarp, × 20. C, seedling, × 1. *e*, endosperm, *r,* three primary ridges of outer surface. *t,* testa. *v,* vittae. *w,* wing.

and half-long (e.g. Offenham) maincrop varieties are commonly used, although shorter 'turnip-rooted' forms exist. A long growing season is required, but in common with many other root vegetables parsnips are now being grown at much higher plant populations than formerly, to produce relatively small uniform roots suitable for pre-packing.

Parsnip roots may be damaged by the disease known as canker (caused by *Itersonilia perplexans* and other organisms), especially on the more peaty soils, and some resistant cultivars such as Avonresister have been developed. Parsnips are rarely attacked by other diseases, and usually only slightly damaged by carrot fly.


### *Apium graveolens* L.   Celery

Celery is a biennial with coarsely bi-pinnate leaves and greenish-white flowers with small entire petals in small, short-stalked umbels. The mericarps are small, 1·5 mm long, 1·75–2·75 million per kg, broadly ovate with entire slender primary ridges; they have the characteristic celery odour, and are sometimes used for flavouring. Leaf-spot disease of celery is seed-borne, and black pycnidia of the causal fungus *Septoria apii-graveolentis* may be present on the mericarps; control is by fungicidal dressing or warm-water treatment.

Wild celery occurs in Britain and over much of Europe as a water-side plant. Cultivated celery (var. *dulce* (Mill.) DC.), which came into use in the seventeenth century, is grown only as a vegetable for

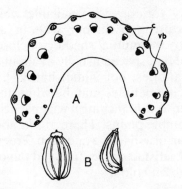

Fig. 33. Celery. A, diagrammatic transverse section of petiole, × 1½. B, dorsal and side views of mericarp, × 10. *c*, collenchyma strands. *vb*, vascular bundle (xylem black, phloem white).

human consumption, both in gardens and, particularly on fen soils, as a field crop. The 'seed' is sown usually in special beds or under glass and later transplanted; direct drilling in the field using pelleted seed or a fluid-drilling technique is sometimes employed, mainly with self-blanching types.

Celery is grown for the thick succulent petioles produced on a very short stem during the first year of growth. The petioles are broad and consist largely of parenchyma, with a series of vascular bundles of varying size extending through it. A strand of collenchyma occurs immediately inside the lower (abaxial) epidermis in association with each bundle; 'stringy' celery is the result of excessive development of these collenchyma strands under unfavourable growing conditions. Ample water is required for the production of high-quality celery. The flavour is due in part to the presence of essential oils produced in oil ducts running through the phloem and parenchyma tissues.

### Range of types

*Winter celery.* The bulk of the main celery crop is derived from cultivars which show some degree of frost hardiness and which require blanching by earthing-up in the latter part of the growing season in order to produce crisp white petioles of the quality desired. Without this blanching the petioles become green, often bitter to the taste, and tough and stringy owing to excessive development of collenchyma. The deep earthing-up required for satisfactory blanching is only practicable on suitable soils, and necessitates very wide spacing (up to 1·5 m) between the rows; in garden practice it is sometimes partially replaced by the use of paper or other collars to exclude light. The blanched petioles are white in the majority of winter celery cultivars (e.g. Giant White), but in some a pink (e.g. Giant Pink) or red (e.g. Giant Red) pigmentation is present.

*Self-blanching celery*. Some celery varieties are available which produce petioles of acceptable texture and flavour without any blanching other than that provided by the mutual shading of closely spaced plants. They are grown on the flat, spaced usually at about 25 cm square, and are mainly early varieties, not winter-hardy, harvested in autumn. At this close spacing petioles may be white (e.g. Avon Pearl) or yellowish (e.g. Golden Self-blanching); wider spacing tends to give green, tough, unmarketable plants. There are in addition a number of cultivars of American origin (e.g. American Green), which, although they give petioles of satisfactory texture and flavour at close spacing, still retain a pale green colour.

*Celeriac*. Celeriac is a very distinct form of the celery species (var. *rapaceum* (Mill.) DC.), grown not for its petioles but as a root vegetable. The short stem and the upper part of the root are much swollen to form a rather irregular globular structure partly above ground level. This is white, with a mild celery-like flavour, and a dry matter content of some 11%. Celeriac is much less commonly grown in Britain than celery, but several cultivars are available, including Alabaster and Globus.

### *Petroselinum crispum* (Mill.) A. W. Hill   **Parsley**

(The nomenclature of parsley is somewhat confused; the E.E.C. classification uses *Petroselinum hortense* Hoffm., and the names *P. sativum* Hoffm. and *Carum petroselinum* Benth. have also been employed.)

Parsley is a perennial of Mediterranean origin with tri-pinnate leaves with wedge-shaped segments, and greenish-yellow flowers in flat umbels, bracts and bracteoles both present. Fruit 2–3 mm, ovoid, with entire primary ridges; mericarps 400 000–700 000 per kg, slow germinating.

#### Range of types

*Sheep's parsley*. A large form with large leaves with broad segments and a woody, usually branched tap-root. Grown occasionally as a grassland herb, either included in a ley mixture, or sown as a constituent of a special herb strip. It is very palatable, but not usually persistent under grazing conditions.

*Curled parsley*. The parsley commonly grown in gardens has smaller leaves with the segments usually much curled and crisped. It is

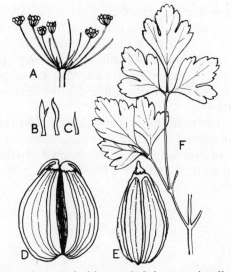

Fig. 34. Sheep's parsley. A, fruiting umbel (part omitted), × ½. B, bracts. C, bracteole, × 1. D, whole fruit, side view, × 10. E, single mericarp, dorsal view, × 10. F, part of leaf, × ½.

used for flavouring and garnishing, and is commonly treated as a biennial. The root resembles that of sheep's parsley; numerous cultivars exist.

*Hamburg parsley*. A very distinct form (var. *tuberosum* (Bernh.) Thell.) in which the unbranched tap-root is swollen and succulent, resembling a small parsnip. It is occasionally grown as a garden root vegetable, and is sometimes known as turnip-rooted parsley; recorded from the sixteenth century in central Europe, but never very commonly grown.

### Other genera

The frequent presence of aromatic compounds in the oil canals of the *Umbelliferae* means that many members of the family can be used for flavouring purposes. A number deserve mention, but none of these is grown on any large scale in Britain.

Those grown for their leaves or other vegetative parts include *Anthriscus cerefolium* (L.) Hoffm., chervil, used in the same way as parsley, and *Foeniculum vulgare* Mill., fennel, of which several cultivated forms exist. In *F. vulgare* var. *dulce* (Mill.) Thell., sweet fennel, the leaves are used for flavouring; *F. vulgare* var. *azoricum* (Mill.) Thell. (var. *dulce* auct non Mill.)., Italian fennel or carasella (often referred to as Florence fennel, although this name apparently applies more properly to sweet fennel), is a distinct form in which the broad sheathing petiole bases are swollen and succulent; it is occasionally grown as a garden vegetable, blanched by earthing-up, and somewhat resembling celery. In *Angelica archangelica* L., angelica, the young green stems and petioles are crystallized with sugar for use as decoration and flavouring.

Those grown for their mericarps include *Pimpinella anisum* L., aniseed; *Carum carvi* L., caraway, which has also been used as a grassland herb in the same way as sheep's parsley; *Coriandrum sativum* L., coriander; *Cuminum cyminum* L., cumin, and *Anethum graveolens* L., dill.

# 6

# SOLANACEAE

*General importance.* The *Solanaceae* includes only one species of outstanding agricultural importance, the potato. Tomatoes and a few other plants are horticultural crops, but many members of the family are poisonous, including a few species found wild in Britain.

*Botanical characters.* Herbs and shrubs, including a few climbers. Leaves exstipulate, simple or pinnately compound, alternate but often partially adnate to the stem so that they diverge from the stem some distance above the node. Inflorescence usually cymose; flowers conspicuous, usually insect-pollinated. Calyx and corolla both of five joined segments, actinomorphic. Stamens five, epipetalous and alternating with corolla segments. Ovary superior, of two joined carpels, surmounted by a single style. Placentation axile, ovules numerous; fruit a capsule or berry, seeds small, endospermic.

### SOLANUM

A large genus, very variable in habit, comprising some 2 000 species of which about 170 produce underground stem tubers. Flowers with rotate corolla, anthers forming a cone around the style and dehiscing by apical pores; fruit a berry.

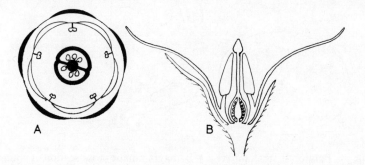

Fig. 35. Floral characters of *Solanaceae*. A, floral diagram. B, vertical section of flower of *Solanum*.

### *Solanum tuberosum* L.   **Potato**

The potato is of South American origin and had been long cultivated there before the discovery of that continent by Europeans. It is not known with certainly in a wild state, but forms one of a large polyploid series of tuber-bearing species, some wild and some locally-cultivated in South and Central America. The poisonous alkaloid solanine is widely distributed in the green parts of the plant, which are therefore not usable for fodder. It is absent from the tubers, except where these are greened by exposure to light.

*Tubers.* The potato is cultivated for its tubers; these are stem structures formed by the enlargement of the tips of underground stems which are variously referred to as rhizomes or stolons. The tubers bear 'eyes'—that is, buds or groups of buds originally formed in the axils of scale-leaves. These scale-leaves are short-lived, and their position is marked in the mature tuber by slight ridges, the enlarged leaf-scars, which form the 'eyebrows'. It is desirable that the tuber should be of regular shape, with shallow eyes, in order to avoid waste in peeling. In most present-day varieties the shape is more or less spherical (*round*) or preferably ovoid (*oval*) or flattened-ovoid (*kidney*); varieties with cylindrical tubers are occasionally grown as salad-potatoes. The eyes are not uniformly distributed but tend to be more closely-spaced towards the apical (rose) end than towards the point (heel) where the tuber was attached to the unthickened part of the rhizome.

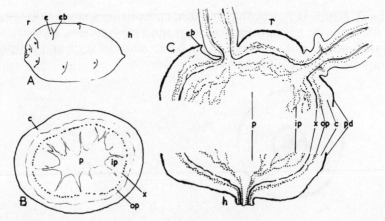

Fig. 36. Potato. A, small tuber, × ½. B, diagrammatic cross-section of young tuber, × 1. C, parts of longitudinal section of sprouted tuber, × 2, xylem strands shown in solid line, phloem dotted. *r*, 'rose end'. *h*, 'heel end'. *e*, eye. *eb*, 'eyebrow'. *p*, pith. *ip*, internal phloem. *x*, xylem. *op*, outer phloem. *c*, cortex. *pd*, periderm.

Anatomically, the tuber shows stem structure. There is little secondary tissue present ; the great increase in diameter is mainly due to the proliferation of the primary tissue, and not to secondary thickening as a result of cambial activity. In transverse section a ring of vascular bundles, which are not clearly separated from one another, surrounds a large irregularly-circular or stellate pith. The amount of xylem is small, and very few lignified cells are present, and the main bulk of the vascular tissue consists of phloem, mainly parenchymatous but with sieve-tubes and companion cells running through it. This phloem is present not only outside the xylem and cambium, but also, and in greater amount, within the xylem, between it and the pith. The internal phloem (also known as medullary, or intra-xylary phloem) occupies a considerable proportion of the cross-sectional area of the tuber. Outside the outer phloem is a rather narrow cortex, surrounded by a well-developed periderm, the thick cork layer forming the skin of the tuber. The original epidermis is only visible in very young tubers. The continuous cork layer of the skin is interrupted at intervals by lenticels, small circular areas in which the cork-cells are loosely arranged, so allowing the exchange of oxygen and carbon dioxide through the otherwise relatively impermeable skin. These lenticels are usually rather inconspicuous, but may increase in size to form raised white dots in tubers grown in partially-waterlogged soil. The main food reserve of the mature tuber consists of starch in the form of large oval starch grains; these are present in all the parenchyma cells, but are more densely packed in the phloem tissue, which therefore appears more opaque than the pith or cortex.

In addition to the normal cork-cambium which produces the tuber skin, further cork cambia may arise in response to wounding. The consequent healing of cut surfaces is made use of in the cutting of large tubers into smaller pieces for planting; this practice is not common in Britain on a field scale, but is sometimes used in gardens. The first result of cutting is the deposit of a layer of suberin on the surface; this is followed by division of the underlying cells to form a new cork cambium parallel to the cut surface, and if conditions are favourable a continuous layer of cork is formed. This depends on the rate of drying out, since if the underlying cells are killed by desiccation before they have produced a cork layer, no healing is possible. If freshly-cut tubers are planted in moist soil, healing is usually satisfactory, but if the soil is dry many pieces may fail to heal, and a 'gappy' plant results. Under such conditions it is better to store the cut pieces in a moist atmosphere for a week or more before planting. Tubers may be cut almost through at the time of boxing, so that the two halves are joined by a narrow strip of tissue. The two cut surfaces thus remain in contact and heal satisfactorily, owing to the fact that they protect each other from drying out. A slight twist is sufficient to separate the two halves at planting time, leaving only a very small area of unhealed surface on each. Dressing the cut surfaces with lime, which was at one time recommended, appears to hinder rather than help the progress of cork formation. Varieties differ in their behaviour; some (e.g. Great Scot) produce cork very readily, while others (e.g. Majestic) heal only slowly.

*Dormancy*. The buds which form the eyes of the tuber are initially dormant, and remain in this condition for a varying period, depending on the variety. In early varieties (see below, p. 130) dormancy is comparatively short, and sprouting (i.e. the growing-out of the buds into shoots) may commence in early autumn in tubers produced during the early summer. In late varieties, dormancy may last until the winter or the following spring; the actual time of sprouting is affected by temperature, and tubers intended for consumption in late spring should be stored under cool (but frost-free) conditions. It has been shown that the period of dormancy can be artificially shortened or lengthened by exposure of the tubers to various chemical substances.

Treatment of dormant tubers with the vapour of ethylene chlorhydrin, for example, results in the breaking of dormancy at an early stage. This is rarely a direct economic value, but has been used in southern U.S.A. to provide a second crop during the year, tubers of the normal early summer crop being replanted after treatment to produce 'new' potatoes in autumn. Gibberellic acid may also be used to break dormancy and give a more uniform crop emergence.

A more generally important use of dormancy-breaking treatments is in testing for the presence of virus diseases. Samples of tubers intended for replanting in the following year are treated soon after lifting, and grown under glass, thus enabling the presence of disease to be detected before the main bulk of tubers is planted.

Treatments which lengthen the period of dormancy are of more direct economic value in that they enable tubers to be kept in good condition for a longer period. Sprouting results in loss both of dry matter and of water from the tubers, and it is often difficult to keep naturally-stored tubers in such a way that they remain in good condition until the following year's crop is available. Tecnazene (2-3-5-6-tetra-chloro-nitrobenzene), originally introduced to stop the spread of dry rot (*Fusarium caeruleum*) in stored tubers, was found to prolong dormancy, as do the growth regulators $\alpha$-naphthyl acetic acid and chlorpropham (with propham). Small amounts of these substances applied at or during storage of tubers result in marked delays in sprouting. In North America the growth regulator maleic hydrazide has also been shown to prolong dormancy following foliar application shortly after flowering. Where tubers are to be used for 'seed' chemical treatment may have to be omitted, modified or curtailed so that sprouting is not delayed beyond the time when it is required.

*Sprouting*. At the end of the period of dormancy, the buds forming the eyes of the tuber grow out if the temperature is high enough (8–13°C). If sprouting starts early, apical dominance is shown, and only a few shoots, near the 'rose' end, develop. If sprouting is delayed by low temperatures (4°C) all the buds tend to grow at the same time, giving numerous sprouts. Shoots produced in the light are short with crowded green leaves; in the dark, shoots are etiolated, long and slender. Etiolated shoots are very liable to damage in handling and planting, and it is therefore desirable to plant as 'seed' either unsprouted tubers or ones which have sprouted in the light, and have shoots not more than about 20 mm long. The number of underground nodes on a stem produced from a tuber previously sprouted in the light will be greater than that on a stem from a tuber planted unsprouted, since in the latter the shoot will develop entirely in the dark. New tubers are produced from stems arising at these underground nodes and, other things being equal, an increase in the number of nodes will tend to result in an increase in number of tubers.

The use of sprouted 'seed' tubers is normal practice with earlies where earliness of yield is all-important; it also allows diseased and damaged tubers to be removed before planting. Although not as widely practised with maincrops there are distinct advantages in using sprouted 'seed' tubers of these; crops bulk earlier and yields are often heavier, especially in years where the growing season is restricted by delayed planting or by premature defoliation by potato blight (*Phytophthora infestans*).

Disadvantages of sprouting include the need for much greater care in the handling of the 'seed' tubers at planting (sprouts may easily be damaged or knocked off), and a greater risk of frost damage as crops emerge earlier. As the whole growth cycle * (emergence, tuber initiation, bulking and maturation) is brought forward, when compared with crops grown from unsprouted tubers, the leaf area duration in a disease-free year, and hence total photosynthesis, may be reduced so

* For a review of the factors affecting sprout growth and subsequent yield, and the concept of 'physiological age' of the potato plant, see Toosey, R. D., *Field Crop Abstracts* **17**, 1964, pp. 162–8, 239–44.

that the final crop yield may be lower than that obtained from unsprouted 'seed' tubers. In normal seasons and with most cultivars the advantages of sprouting outweigh the disadvantages.

In order to facilitate planting with fully automatic planters 'mini-chitted' tubers may be used. Here the 'seed' coming from bulk store is 'cured' at about 13°C for about ten days to cure any damage caused by handling and to check gangrene (*Phoma exigua* var. *foveata*). During this time the eyes begin to sprout and their growth is then retarded by storage at 4°C until about three weeks before planting, when the temperature is allowed to rise to 7–10°C to start the sprouts growing again. Sprout growth of some 3 mm only is much less than with fully sprouted 'seed' tubers, and mini-chitted tubers will withstand bulk handling; special chitting trays and artificial lighting are unnecessary, and mini-chitting can be carried out in sacks or deep boxes, but on the other hand refrigerating equipment is needed to maintain the required low temperature. The field performance of mini-chitted tubers is usually intermediate between that of fully sprouted and unsprouted 'seed' tubers.

*Development of the planted tuber.* The sprouts develop at the expense of the starch and other materials present in the tuber, and form stems which bear alternately-arranged leaves. Below ground, these leaves are small and scale-like, and the stems round and colourless, with numerous adventitious roots arising at their nodes. Above ground the stems are square or triangular in section, with conspicuous, sometimes wavy wings at the corners, formed by the decurrent bases of the petioles. The leaves are large and pinnately compound, with a single terminal leaflet and a number of pairs of large, stalked, lateral leaflets. Between these large primary leaflets are a number of almost sessile small secondary leaflets or folioles, and still smaller tertiary leaflets are usually present on the stalks of the primary leaflets. The leaflets are entire, usually ovate with cordate base, but the size and shape varies considerably in different varieties, as does the length and degree of erectness of the stems.

Buds in the axils of underground scale-leaves grow out to form horizontally-directed rhizomes, which bear scale-leaves and adventitious roots, but do not usually branch. The tip only of each of these slender rhizomes acts as a food-storing organ, and swells to form a tuber, which is thus connected to the upright stem by the unswollen part of the rhizome. This is short in the majority of potato varieties, but may reach a length of several feet in the related wild species; long rhizomes are usually associated with low tuber yields.

Meanwhile, the food material stored in the planted 'seed' tuber has

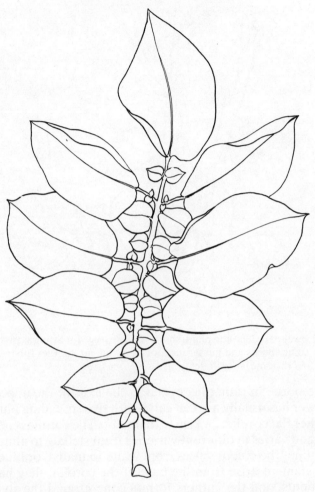

Fig. 37. Potato leaf (Majestic), $\times \frac{1}{2}$.

been used up by the developing stems, and is not replaced. The 'seed' tuber thus dies, and the stems which arose from different eyes and which were previously only connected by the tissue of the 'seed' tuber become entirely separated from one another. The apparent potato plant, derived from a single 'seed' tuber, is therefore no longer strictly one plant, but a colony of independent unconnected plants growing close together. This lack of organic connection between the different stems in the later stages of growth is of importance in limiting the spread of virus diseases in cases where infection takes place late in the year.

As the aerial stems approach their full size they produce flowers,

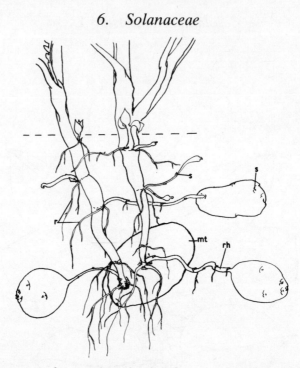

Fig. 38. Lower part of potato plant, shortly before stage for lifting as first early, × ½. Simplified, some shoots and numerous roots omitted. *mt*, mother tuber, *r*, adventitious root. *rh*, rhizome. *s*, scale leaf.

which are borne in rather lax cymes in the axils of the upper leaves. Each flower has usually a green calyx with five spreading blunt teeth, and a rather flat corolla, 2–3 cm in diameter. The corolla is not deeply divided, and varies in different varieties from stellate to almost circular in outline; the colour varies from white to mauve or blue-purple. The five stamens arise from the base of the corolla; they have short stiff filaments, and the anthers form a cone around the style. Since potatoes are propagated vegetatively (except for plant-breeding purposes) the flower is of no distinct economic importance, and many varieties, selected for their desirable vegetative characters, rarely flower or have distorted and partially or wholly sterile flowers. In such varieties the anthers are often malformed or partially aborted and form an irregular, twisted, pale yellow cone, producing little or no viable pollen. In fertile varieties the anthers are usually broader and deep orange in colour. The ovary is globular, with a slender style of variable length, bearing a capitate or slightly-notched stigma.

In many varieties the flowers are shed at an early stage, but in those in which they are retained, and in which pollination takes place, the ovary develops to form a globular berry, two-chambered and with

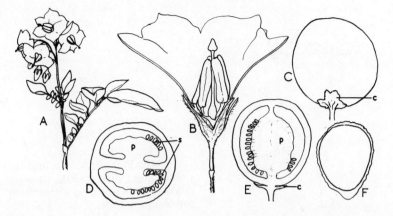

Fig. 39. Potato flower and fruit. A, inflorescence, × ⅓. B, flower, with part of corolla removed to show style and stamens, × 2. C, mature fruit, × 1. D, transverse, and E, longitudinal sections of fruit, × 1. F, seed, × 10. *c*, calyx. *p*, placenta, *s*, seeds (a few only shown).

large, fleshy placentae on which numerous seeds are borne. The ripe seeds are pointed-oval in shape, about 2 mm long, flattened and slightly winged, pale brown in colour and rough or slightly hairy in appearance, owing to the partial breakdown of the epidermal cells. The embryo is curved and embedded in the endosperm.

These true seeds are used only when it is desired to produce a new variety and not for the production of a crop. Germination is epigeal and the cotyledons become green, ovate and leaf-like. The epicotyl elongates quickly and bears a succession of alternately-arranged simple ovate leaves. While the plant is still small, buds in the axils of the cotyledons and lower leaves grow out as slender stems which turn downwards and enter the soil; there their tips enlarge to form tubers. These first-year tubers usually reach a diameter of only 1–3 cm; in the breeding of new varieties they are lifted, stored, and replanted the following year, when they give rise to plants bearing rather larger tubers. These are treated in the same way and full-sized tubers are produced in the third or fourth year.

*Yield*. Potatoes require a high level of soil fertility, and heavy manuring is necessary for high yields. For satisfactory growth of the tubers it is also necessary that cultivations shall be such as to give as little consolidation of the soil as possible, and to prevent weed competition in the early stages. Given these conditions, yield will depend largely on the amount of photosynthesis which the plants are able to

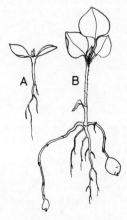

Fig. 40. Potato. A, seedling. B, young plant from seed, showing origin of tuber-bearing stems from cotyledonary node, × 1.

carry out. In the early stages the plants are not self-supporting, but are dependent on the food reserves in the planted tuber; this food reserve is, of course, much greater in amount than in a crop grown from the true seed. Weight of tubers planted per acre has therefore an important effect on final yield, and the yield per acre increases with increase of 'seed' rate up to at least 5 tonnes of 'seed' per hectare. The relationship between ware yield and 'seed' rate is a complex one and is affected by variety, 'seed' size, number of sprouts per tuber, row width, planting distance within rows and availability of nutrients and water. 'Seed' rates of between 2·5 and 3·75 t/ha are commonly used but rates of up to 7·5 t/ha have been used on very fertile irrigated soils, or where high yields of small tubers are required for canning. Providing that the total weight of tubers planted per hectare is the same the size of 'seed' has little effect on total yield, but may have a significant effect on the proportion of the crop reaching ware size. Tubers of about 55–60 g mass are usually preferred for planting; very large 'seed' may give a crop of the same total weight per hectare, but consisting of a larger number of smaller tubers. Crops can be grown from small chips including an eye, or even from detached eyes, but the yield is normally low.

The foliage produced at the expense of the planted food reserves carries out photosynthesis; the products of photosynthesis are used partly in the production of further leaves, partly in the development of tubers. Varieties which produce tubers early are thus not able to develop as large a leaf area as varieties in which tuber-production is delayed; the total amount of photosynthesis, and consequently the total yield, is thus lower in early varieties. Any factors which tend to decrease the area or duration of the leaves will, of course, tend to decrease yield; thus blight is of outstanding importance, not only because of the direct tuber damage it may cause, but also owing to the premature defoliation it produces, which markedly lowers the yield even when the tubers escape infection. Flowering has a much less marked effect on yield than it has in biennial crops, but does make some demands on the available food materials. It has been shown experimentally* that in varieties which flower and fruit abundantly, tuber yield is increased by up to about 20% if fruiting is artifically prevented, and by up to about 40% if both flowering and fruiting are stopped.

*Quality*. Since potatoes are grown primarily for human consumption, quality is of very great importance. A large number of factors contribute to quality and requirements differ in different countries and in different areas. Even-shaped tubers of even size, with shallow eyes, are always preferred; cracked, hollow tubers are undesirable. Second growth, or 'super-tuberation', in which partially mature tubers grow out to form attached daughter-tubers, is a serious defect. Skin colour is of no direct importance, but purple-skinned varieties are not popular in England; there is a demand for tubers with red coloration around the eyes, since this is associated in the consumer's mind, but not necessarily in fact, with the good quality of the variety King Edward. 'Flesh' colour is regarded as important, with a marked preference in Britain for white-fleshed varieties. The few yellow-fleshed varieties grown include in the 1970s the first-early Duke of York, mainly in private garden use, and the second-early Wilja and maincrops Désirée and Record; the latter three are introductions from continental Europe where yellow-fleshed varieties are widely grown.

For normal domestic use dry-matter content is not in itself important; the usual percentage is about 23 for the whole tuber, but it is not uniformly distributed, and the pith may contain less than 10%. Cooking quality may, however, be better in tubers with a high dry-matter content, although this character is difficult to assess precisely. For most purposes British taste calls for tubers which are floury when cooked, but firmer flesh is desirable in tubers for frying, and also in earlies to be eaten as 'new' potatoes. Where tubers are used for processing (about 20% of crop in mid-1970s) dry-matter content may be very important. High dry-matter is required for potato chips, potato crisps and dehydrated potato (instant mash), and low dry-matter in 'waxy' tubers canned whole as 'new' potatoes. Potatoes for crisps or chips should in addition have a low content of reducing

* Bartholdi, W. L., *Univ. of Minnesota Agr. Expt. Station Tech. Bulletin* **150**, 1942.

sugars (about 0·18% of fresh weight is considered ideal), as too much sugar produces a dark brown colour in the finished product. Storage of tubers at temperatures below 4°C results in an increase in sugar content.

Blackening of tubers after cooking is a serious defect associated partly with variety and partly with soil conditions. Varieties with a high chlorogenic acid and low citric acid content such as Home Guard and Majestic commonly show this type of discoloration, whereas varieties in which the levels of the two acids are reversed are usually free from it; King Edward and Maris Piper are examples. A different type of discoloration is the enzymatic browning of raw tubers after peeling, which is shown by some varieties, and in particular by Majestic. This is caused by the enzyme polyphenyl oxidase acting on tyrosine to produce the dark-coloured melanin. The reaction takes place only in broken cells in the presence of oxygen, and immersion of the peeled potatoes in water will prevent discoloration as the contents are leached out of the damaged cells and the oxygen supply is reduced.

A number of other tuber characters also have an important effect on the value of a variety; among these are resistance to mechanical damage and resistance to various storage diseases. Varieties which bruise easily during lifting and grading may show various types of internal discoloration in apparently sound tubers. Susceptibility to diseases which spread during storage (such as dry rot) is particularly important in the tubers kept for 'seed', which may appear healthy but may, if infected, fail to produce shoots.

### Origin and range of types

A large range of tuber-bearing *Solanum* species occurs in South and Central America, extending as far north as Mexico. They form a polyploid series based on a haploid chromosome number $x = 12$. The interrelations of the different species have not been fully worked out, but some of the polyploids may well be amphidiploids derived from natural crosses between species. The most recent evidence indicates that the cultivated tetraploids (which include the main cultivated species) arose in the high plateau regions of Peru, Ecuador and Bolivia, probably from *Solanum stenotomum*, a diploid which has apparently been in cultivation since the third millenium B.C. or earlier. The Peruvian type of cultivated tetraploid, *S. andigena*, is adapted to produce tubers under the short days of equatorial regions. As cultivation spread southwards to other parts of South America natural adaptation to other environmental conditions took place giving rise in Chile and the off-shore island of Chiloe to forms morphologically similar, but producing tubers under the long day conditions of temperate summers. Although these Chilean forms are essentially similar to the modern European potato and are included with them in *S. tuberosum*, the historical evidence indicates that the first potatoes introduced into Europe (in Spain *c*. 1570) were of the Peruvian (*S. andigena*) type, which produced tubers only in the short days of late autumn. The first introduction into Britain, also of Peruvian type, occurred in about 1590, and appears to have been independent of the earlier Spanish introduction.

It was not until some 200 years later that potatoes became significant as a crop in Europe. By this time varieties adapted to long days had emerged as a result of selection for earliness and yield of plants grown from true seed; natural selection under European conditions doubtless played a part in this change.

There is no firm evidence to suggest that further introductions, if in fact there were any during this period, played any part in the development of the European potato. It would appear that the sixteenth-century introductions were highly heterozygous and provided a large source of genetic variation. The development of the European potato appears therefore to have occurred independently in essentially the same manner as that of the Chilean potato. Apart

**Table 2. Examples of tuber-bearing species of *Solanum*, wild and cultivated, to illustrate the polyploid series** (Note that *S. andigena* is sometimes written *andigenum*.)

| | WILD | CULTIVATED |
|---|---|---|
| **DIPLOID**<br>2x = 24 | *S. sparsipilum* (Bitt.) Juz. et Buk. | *S. stenotomum* Juz. et Buk. |
| **TRIPLOID**<br>3x = 36 | *S. commersonii* Dun. | *S. juzepczukii* Juz. et Buk. |
| **TETRAPLOID**<br>4x = 48 | *S. acaule* Bitt. | *S. tuberosum* L.<br>*S. andigena* Juz. et Buk. |
| **PENTAPLOID**<br>5x = 60 | *S. curtilobum* Juz. et Buk. | |
| **HEXAPLOID**<br>6x = 72 | *S. demissum* Lindl. | |

from the differénces in day-length requirements the differences between the Peruvian and the European (and Chilean) forms are so
small that some authorities classify them as subspecies of the one
species *S. tuberosum*, viz. subsp. *andigena* and subsp. *tuberosum*.

Descendants of a Chilean variety introduced into North America
in about 1850 under the name of Rough Purple Chile may have been
important in the development of varieties, particularly early varieties, in the latter part of the nineteenth century. The first interspecific crosses were made at about this time, but it was not until
towards the middle of the twentieth century that wild species were
introduced into breeding programmes on any scale in attempts to
improve resistance to disease, eelworms and frost.

Whatever their origin, the cultivated potatoes of Europe and other
temperate regions now comprise a series of long-day forms. The
short-day forms maturing tubers only in very late autumn have been
discarded, but those now grown show a continuous range of maturity
dates from early summer to early autumn. They can be arbitrarily
divided into first earlies, second earlies, early maincrops and maincrops; but it should be emphasized that this classification is based on
agricultural rather than botanical characters. Thus first earlies
include not only such varieties as the almost obsolete May Queen,
which produces its full tuber-yield early, but also the much more
commonly grown Arran Pilot, the full yield of which is not reached
until the second early period. This latter variety shows, however,
early 'bulking-up', and a sufficient proportion of the yield is produced
early in the season to make it economic to lift it as a first early. All
varieties other than maincrops are normally consumed soon after
harvesting, and keeping quality is not important except in relation to
'seed'. Maincrops, which commonly produce the highest yield per
hectare, may need, however, to be kept until the following June, and
long dormancy and good keeping quality is therefore required.

### Cultivars

Since potatoes are vegetatively propagated by means of their tubers,
the varieties are strictly clones, and each variety (cultivar) is derived
from one single true seed. Occasionally somatic mutation takes place,
and the mutated forms may be selected and multiplied up so that a
second variety is produced, derived from the same original seedling.
Thus King Edward, with tubers coloured red around the eyes only,
has given rise to the all-red but otherwise similar Red King. Usually
such mutation affects only certain tissues, and the resultant plant is a
chimaera, thus if the eyes are removed from a tuber of the russet-

skinned Golden Wonder, and adventitious shoots arise from the deeper layers of the tuber, the tubers which these shoots produce are smooth-skinned, and indistinguishable from those of Langworthy, from which variety Golden Wonder presumably arose.

The requirements of a good potato variety are so many and varied that no very clear picture of the development and interrelations of varieties can be presented. Out of the very large number of new varieties produced, very few prove to be of sufficient all-round excellence to displace the existing older varieties. Although much progress has been made in the development of disease-resistant varieties by crosses with other *Solanum* species, the majority of such varieties have so far suffered from defects in their field characters or quality which have prevented their extensive use. The slow rate of introduction of new cultivars superior to existing ones is well shown by the fact that the most widely grown cultivar in Britain in 1954 was Majestic (50% of maincrop area) introduced in 1911; this was followed in popularity by King Edward (25% of maincrop area) introduced in 1902. In the same year Arran Pilot (introduced 1930) occupied 50% of the first-early area, followed by Home Guard (1943) with 25%. Since 1950 a considerable number of new cultivars of some merit have been introduced as a result of the intensification of breeding programmes, and most of the older varieties are no longer widely grown. Thus on the N.I.A.B. Recommended List of Potato Varieties for 1978/9 only about one-third of the total number of varieties named were introduced prior to 1955, and about one-fifth had been introduced since 1970.

*First earlies.* (Normally lifted at an immature stage when haulm still green.) In 1977 Arran Pilot (introduced 1930), early bulking and high yielding but of rather low quality, for long the most widely grown variety in this group, occupied less than 3% of the first-early area in Britain. The most popular varieties were Pentland Javelin (1968), slightly later maturing but high yielding and of good quality, with 25% of the area; Ulster Sceptre (1962), very early and high yielding, with 23%; and Home Guard (1943), which had retained its popularity at 23%, partly on account of its good quality for processing as potato crisps when left to maturity. All these varieties have oval to long-oval tubers, shallow eyes, white skin and white flesh.

*Second earlies.* (Lifted at immature stage to follow on from first earlies, but may be left to mature as early maincrop if market glutted.) By far the most widely grown variety in Britain in 1977 was Maris Peer (1962) which occupied 64% of the area grown. This variety has short-oval, medium to small tubers with white skin and white flesh. When grown at high 'seed' rates and lifted early it produces tubers of a size and quality very suitable for canning as 'new potatoes'. Other varieties include Craigs Alliance (1948), white skin and white flesh, and Red Craigs Royal (introduced 1957 as a fully red-skinned mutant from the particoloured Craigs Royal (1948) which it has now largely replaced); each of these good quality varieties accounted for about 10% of the second-early area in 1977. A more recent introduction is the high yielding yellow-fleshed Dutch cultivar Wilja (1974) occupying a further 9%. First and second earlies together account for between one-fifth and one-quarter of the total potato area in Britain.

*Maincrops.* (Normally lifted at full maturity and stored in clamps or special stores until required for sale.) For many years Majestic (1911), with oval to long-oval white-skinned and white-fleshed tubers, was, although not outstanding in any particular character, the most widely grown variety on account of its general level of yield and quality. Since about 1960 several Majestic replacements have been introduced and in 1977 Majestic occupied less than 2% of the maincrop area. The most popular variety was Pentland Crown (1958), very high yielding with short-oval tubers with white skin and flesh, occupying some 27% of the area. Pentland Dell (1960), similar but earlier and with longer oval tubers, showed considerable promise when first introduced, since it was then resistant to potato blight. It proved however to be susceptible to new races of the fungus, and by 1977, when it occupied some 6% of the area, was classed as very susceptible to tuber blight. Other Majestic replacements include Maris Piper (1963) with some 17%; it is resistant to one form of the potato cyst eelworm (*Globodera rostochiensis* pathotype A) but very susceptible to drought and to slug damage. King Edward (1902), with oval to long-oval particoloured cream-fleshed tubers, retained its reputation as the best quality potato, and in spite of its relatively low yield, marked susceptibility to blight and lack of immunity to wart disease, still occupied 11% of the maincrop area in 1977. Other recommended varieties include Désirée (1962) with red skin and yellow flesh, high yielding and of good quality, of Dutch origin and occupying 14% of the area; and Record (1932), also Dutch and with yellow flesh, which has a high dry-matter content and is grown in Britain almost exclusively for processing as potato crisps, and accounted for 8% of the maincrop area.

*Private garden varieties.* Considerable quantities of potatoes are grown in private gardens and allotments and the varieties used include many which are obsolete or little used for commercial production. They include amongst first earlies Duke of York (1891, yellow-fleshed and not immune to wart disease), Epicure (1897, still in commercial use in Scotland), and Sharpes Express (1901); amongst second earlies Catriona (1920); maincrops Dunbar Standard (1936), Golden Wonder (1906), and Kerrs Pink (1917, also still grown commercially in Scotland).

The U.K. National List of Potato Varieties (September 1977) gives 106 cultivars; of these 22 are included in the N.I.A.B. Recommended List 1978/9.

*Solanum melongena* L., **aubergine**, is a non-tuberous species cultivated for its edible fruit. It is a perennial, cultivated as an annual, with large simple ovate lobed leaves, purple flowers, and usually elongated deep purple fruits (ovoid white in one non-edible form, whence the alternative name, egg-plant). Fruit eaten cooked, firm fleshed almost solid, so that seeds appear embedded in parenchyma. Of Indian origin, introduced into Europe in the fifteenth century, and occasionally cultivated under glass in Britain; F.1 cultivars have been developed.

### *Lycopersicon esculentum* Mill.   **Tomato**

(The E.E.C. classification uses the name *Solanum lycopersicum* L. with *Lycopersicum esculentum* Mill. as a synonym.)

An annual, closely related to Solanum, leaves somewhat resembling those of the potato, but with leaflets pinnately cut or divided, secondary leaflets common, glandular-hairy; stem weak, sprawling; flowers in apparently leaf-opposed cymes, corolla deeply lobed, yellow, anthers forming cone but with longitudinal not porose dehiscence. Grown for the succulent edible fruit, relatively thick walled and with pulpy outgrowths from the placentae around the seeds. Seeds similar in form to those of the potato.

Native in north-western South America, in a form (var. *cerasiforme* (Dun.) Alef.) with small cherry-like fruits and pentamerous flowers with short calyx teeth and two carpels. Introduced to Europe early in the sixteenth century, now widely grown in a large range of cultivated self-pollinating varieties, with large or small, globular, elongated or pear-shaped red or yellow fruits. Tomatoes were originally largely cross-pollinated by insects, the flowers having the style long-exserted through the anther cone before the pollen was shed. In modern cultivars the style is of the same length as, or shorter than, the anther cone, and self-pollination is normal. Commercial cultivars in Britain are mainly hexamerous with red globular bilocular fruits. Varieties exist in which the fruits are multi-locular and much more solid-fleshed owing to the numerous septa and enlarged placentae; seed production in these cultivars is often much lower than in those with bilocular fruits.

Tomatoes were not important in Britain until the end of the nineteenth century; they are now extensively grown, mainly under glass, but some outdoor varieties (not frost-hardy) exist, including some dwarfs. Numerous F.1 cultivars have been developed. Now one of the most important 'vegetables' (botanically a fruit) on a world scale; eaten raw, cooked, or processed as soup, sauces, purées, pickled or canned. Oil from the seeds extracted during processing is used for margarine and soap manufacture.

The close-related *L. pimpinellifolium* Dun (*L. racemigerum* Lange), with glabrous leaves and red-currant-like fruits, crosses readily with tomato and has been used in the production of disease-resistant forms.

### *Capsicum annum* L.   Sweet Peppers, Chillies

*Capsicum annum* is an annual with a much branched stem; leaves simple, entire, lanceolate to ovate with acuminate tips; flowers usually solitary, terminal with five- and six-lobed white corolla; fruit originally two-celled but may become one-celled at maturity, pericarp firm, fruit more or less hollow when mature, very variable in size, shape, colour and pungency. Central American in origin, introduced to Europe early in the sixteenth century, and now widely grown in tropics and subtropics.

Sweet peppers (bell peppers, pimientos) normally have large conical or blunt-ended somewhat inflated fruits, and may be used green or when ripe (red or yellow). The pericarp is thick and mild-tasting. Widely cultivated in warm climates, some production under glass in Britain; F.1 cultivars are common. Paprika is prepared from similar forms and contains the red pigment capsanthin.

Other forms with fruit of much greater pungency owing to their higher content of the pungent principle capsicin are grown as chillies, red pepper, Cayenne pepper, etc. The fruits may be long or short, pendulous or erect, red or yellow at maturity; some forms are occasionally grown as ornamentals. *C. frutescens* L., bird chilli, is a similar but somewhat shrubby perennial species with small clustered erect extremely pungent fruits, used in the same way. These pungent forms are much used (especially in hot countries) as spices, either on their own or as mixtures with other spices, e.g. in curry powder; they may be good sources of vitamins A and C. (The condiments black and white pepper are quite distinct from these *Capsicum* peppers, and are derived from the unrelated *Piper nigrum* L. of south-eastern Asia.)

## *Nicotiana tabacum* L.    Tobacco

Tobacco is a tall annual, with pink-tinged short tubular flowers and the fruit a capsule with numerous very small seeds, grown for its large ovate-spatulate leaves, which contain nicotine and are used after curing and fermenting. A large range of forms are grown, producing a variety of leaf types suitable for particular kinds of manufactured tobacco; soil conditions and method of curing can have marked effects on quality. Of South and Central American origin, now widely grown in warm temperate and subtropical countries; can be grown in southern England but not economic on a commercial scale. Nicotine, used as an insecticide or oxidized to nicotinic acid for use in vitamin preparations, is extracted from tobacco waste. The related *N. rustica* L. has also been used for nicotine and citric acid production.

# 7

# LEGUMINOSAE

The *Leguminosae* is a very large family, divided into three sub-families of which only one, the *Papilionatae,* is of importance in temperate regions. It is to this sub-family (sometimes treated as a separate family, the *Papilionaceae*) that the following particulars apply:

*Habit.* Very variable, trees, shrubs and herbs. The latter include annuals and perennials, and may be erect or climbing, tufted or creeping.

*Leaves.* Alternate, stipulate, usually compound, often with tendrils.

*Flowers.* Conspicuous, usually insect-pollinated, in racemose inflorescences. The flowers are of a characteristic and highly-specialized zygomorphic type, called papilionate from its fancied resemblance to a butterfly in the large-flowered species. The calyx is composed of five joined sepals; the five petals vary in shape. The posterior one is often large and erect, forming the *standard* (vexillum) and partially overlapping the smaller lateral petals or *wings* (alae) which lie one on each side of the *keel* (carina) formed by the fusion of the two anterior petals. Within the keel lie the ten stamens; all ten may be joined by their filaments to form a tube surrounding the ovary (*monadelphous*), or the posterior one may be free, the other nine forming an incomplete tube (*diadelphous*). The gynaecium is of one carpel, usually with numerous anatropous ovules, the style emerging from the open end of the stamen-tube and curving upwards between the anthers.

Pollination is usually by bees. Nectar is secreted in diadelphous forms within the base of the stamen-tube, and can be reached only by long-'tongued' insects through the slits between the free stamen and those adjacent to it. In some *Leguminosae,* e.g. *Vicia,* projections of wing and keel petals interlock so that an insect alighting on the wings depresses the keel and exposes the stigma and anthers. In others, e.g.

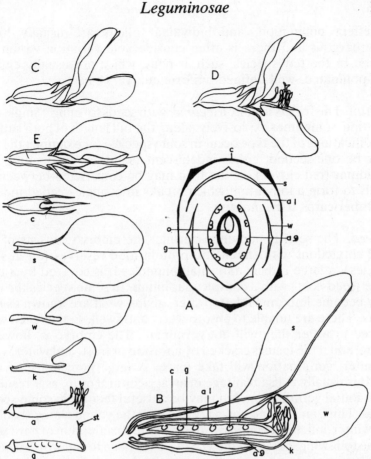

Fig. 41. Floral structure of *Leguminosae*. A, floral diagram. B, vertical section of broad bean flower, × 2. C, flower in normal position. D, with wings and keel depressed, both in side view. E, ventral view and separate parts of flower, all × 1. *a*, androecium. *a*1, separate stamen. *a*9, joined stamens. *c*, calyx. *g*, carpel. *k*, keel. *o*, ovule. *s*, standard. *sg*, stigma. *w*, wing.

lucerne, the ovary and staminal tube tend to curve upwards and are held down by the rigid keel; dehiscence of the anthers takes place in the bud stage, and the stigma is thus covered with pollen. The stigma is not, however, receptive unless the flower is 'tripped'—that is, the ovary is released from the keel so that it springs sharply upwards and the stigmatic surface is ruptured. Tripping by bees usually results in cross-pollination; in some plants self-tripping may occur and give rise to selfing. In the majority of *Leguminosae* cross-pollination is usual, and many species are almost completely self-sterile. In such plants seed production is largely dependent on an adequate number of bees to effect pollination, and individual plants are usually highly heterozygous and there is often considerable variation within cultivars. In the few species, such as peas, which are usually entirely self-pollinated, well defined uniform cultivars exist.

*Fruit.* The fruit is usually a *legume,* with seeds forming a single row, splitting, sometimes explosively, along the full length of both sutures. Modifications of the type occur in some species; for example, the fruit may be one-seeded, either indehiscent (sainfoin) or opening as a pyxidium (red clover), or the fruit may be constricted between the seeds to form a lomentum which breaks into single-seeded indehiscent mericarps (*Ornithopus*).

*Seed.* The seed is non-endospermic; the embryo consists of two oval cotyledons with starch and protein food reserves (oil in some species), a large radicle and small plumule. It is covered by a thick testa (seed-coat) with a conspicuous hilum. In some species the testa may become impermeable to water, giving what are known as *hard seeds.* These are unable to absorb water and swell, so that even when placed in water they will not germinate. The embryo is, however, living, and if the testa is cracked by abrasion or frost, and water is able to enter, germination will take place. A large percentage of hard seeds is usually a disadvantage in an agricultural crop, as it results in poor initial germination, but may be a useful feature in some special cases. Thus, an autumn-sown ley in which the young clover seedlings are winter-killed may improve owing to the germination of hard seeds in the following spring. In the U.S.A. it has been found worth while to develop special strains of the annual crimson clover with a high percentage of hard seed, which can persist in the soil and give a cover of clover after an intervening cereal crop, without any further sowing.

*Root-nodules.* The root system of leguminous plants is typically a branching tap-root. It differs from that of the majority of families of flowering plants in bearing root-nodules. These are lateral outgrowths of the roots, mainly parenchymatous, but with a vascular supply connecting up with the vascular system of the roots. They are caused by the presence of particular bacteria, *Rhizobium radicicola,* which enter through the root-hairs. They multiply in the root-hair and form an *infection-thread,* a slender filament of bacterial cells which becomes surrounded by a tube-like sheath secreted by the root cells. This infection-thread penetrates through the cortical cells, and in the inner cortex the sheath breaks down, setting free the bacteria in the host cells. Here they multiply and stimulate the host cells to divide, thus causing proliferation of the inner cortex and the development of a nodule.

*Rhizobium radicicola* is capable of utilizing the free nitrogen of the air, and if conditions are favourable a state of *symbiosis* is set up in which the bacteria supply nitrogenous compounds to the plant while the leguminous plant provides the bacteria with carbohydrates. If satisfactory symbiosis occurs, the leguminous host plant becomes independent of combined nitrogen in the soil. This depends on soil conditions; for example, nitrogen fixation does not take place in waterlogged soils or in the absence of boron. It is also dependent on the strain of bacterium.

*Rhizobium radicicola* is a variable species, consisting of a number of distinct forms, sometimes treated as separate species, capable of forming nodules on certain leguminous genera, but not on others. Thus one form (*R. leguminosarum,* if it is treated as a species) produces nodules on species of *Pisum* and *Vicia,* another (*R. trifolii*) on clovers, and so on. Within each of these groups numerous strains exist, some effective nitrogen fixers, others ineffective; ineffective strains may behave as parasites, and harm rather than assist the host plant. The bacteria set free by the decay of the nodules remain viable in the soil for some years, and a field which has carried a satisfactorily nodulated crop of a particular leguminous species will give satisfactory results with the same species in later years. If, however, a new leguminous species is sown, which had not previously grown on that particular land, results may be poor, owing to the absence of the appropriate strain of *Rhizobium.* Inoculation with a suitable strain is then necessary. This may be done either by scattering over the land soil taken from a field where the legume is known to make good growth, or more conveniently by the use of a pure culture of the bacterium. Cultures can be kept growing on agar media, and transferred to a skim-milk medium, which is then mixed with the seed, which is dried, away from sunlight, and sown as soon as possible. If this is done, bacteria of the correct strain are present on the seed-coat at the time of germination, and infection of the young roots readily takes place. In Britain lucerne is the only crop which is commonly inoculated in this way; it can usually be assumed that the appropriate strains for clovers and the other common legumes are already present in all agricultural land.

*Agricultural value.* The *Leguminosae* are of outstanding agricultural value because of their specialized form of nutrition. The nitrogen fixed by the *Rhizobium* cells in the root-nodules is utilized by the leguminous plants, which are thus not dependent, as are almost all other plants, on nitrogen compounds present in the soil. They are valuable in themselves, therefore, as high-protein food plants, and also for their effect on other plants, since some of the nitrogen becomes available to nearby non-leguminous plants by decay of the nodules or, under some conditions, by excretion of nitrogenous compounds from the nodules. The amount of nitrogen fixed in the root-nodules will vary with the species and variety of plant, with soil conditions and with season. Estimates of total nitrogen fixation during the growing season range from up to 80 kg/ha in peas to over 600 kg/ha in perennial clovers.*

Members of the family are used mainly in two ways: the large-seeded forms such as peas and beans are grown as arable crops for their high-protein seed reserves (pulse crops), while the smaller-seeded forms, such as clovers and lucerne, are grown as forage crops, either alone or in mixture with grasses. Other methods of utilization include the use of immature pods as vegetables (runner and French beans), of quick-growing leafy forms as green manure, and of large-flowered forms as ornamentals. Some tropical and warm-temperate species are used as oil-seeds, for producing insecticides and fish poisons, for drugs and for resins and gums.

## HERBAGE LEGUMES

### TRIFOLIUM (TRUE CLOVERS)

Herbaceous plants, annual or perennial, leaves trifoliate, leaflets at least slightly toothed, with obtuse or emarginate tips. Flowers small in short, crowded racemes, standard only slightly diverging from other petals, stamens diadelphous. Pods one- to several-seeded, short, straight, dehiscent. First leaf of seedling simple.

A rather large genus including several species of outstanding importance as herbage plants in temperate regions.

### *Trifolium pratense* L.   **Red Clover**

*Distinguishing characters.* Hairy perennial, with stout tap-root, not creeping. Leaves large, leaflets almost entire, usually with horseshoe-shaped leaf mark. Stipules large, often red-veined, broad and narrowing abruptly to a short, slender point. Flowers densely crowded in ovoid racemes subtended by leaf-like bract. Calyx hairy

---

* For further information see Nutman, P. S. (ed.), *Symbiotic Nitrogen Fixation.* 1975. Cambridge University Press.

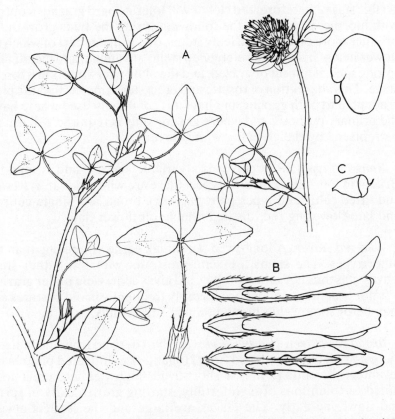

Fig. 42. Red clover. A, vegetative and flowering shoots and detached leaf, × ½. B, side, dorsal and ventral views of single flower, × 3½. C, fruit, × 3½. D, seed, × 10.

with narrow teeth. Petals purplish-red (occasionally white), forming a long tube at base, pollinated mainly by humble-bees. Mainly self-sterile. Pod very short, single-seeded, opening transversely as a pyxidium. Seed 2 mm, 550 000 per kg, oval, but with radicle forming a conspicuous lobe ⅓–½ length of cotyledons. Testa shiny, pale yellow with variable purple shading. Seed harvested under wet conditions may lose most of its purple pigment by weathering. Old seed darkens and becomes brown. Proportion of hard seeds usually low. Seedling with short-stalked, oval cotyledons, *c.* 5 mm long. First leaf simple, rounded, sub-cordate, 4 mm long by 6 mm broad, hairy.

*Growth habit.* The seedling develops a tap-root, and during the first year's growth the stem remains short so that a rosette of leaves is produced, with the axillary and terminal buds at or near ground-level.

At this stage, therefore, red clover will tolerate hard grazing (contrast with lucerne, p. 158). In the following spring the buds grow out to produce more or less erect, leafy stems, on the upper part of which the flower heads arise. These elongated stems are annual and die off after setting seed, if not cut or grazed, and the plant overwinters in a rosette stage. This alternation of rosette and erect stages means that the plant is adapted to both grazing and hay; it is commonly used where hay is the primary purpose, and where grazing only is required will usually be replaced by the creeping white clover.

*Range of types.* Red clover is a variable species, and is found in Britain in four rather distinct types; these are wild red (var. *sylvestre*) and three cultivated types (var. *sativum*), broad red, single-cut red, and late-flowering red, distinguishable on flower date.

*Wild red clover.* A long-lived perennial, earlier-flowering than the other types. The stems are semi-prostrate, with few rather small leaves; it is thus not suited for hay and gives a low yield under grazing. It is not in cultivation, but occurs fairly frequently in old pastures and may be of some value in maintaining soil fertility.

*Broad red clover. (Early red clover.)* Not derived from the indigenous wild red, but introduced from Holland as a cultivated plant in the seventeenth century. It is a short-lived, high-yielding, erect form suited to conditions of high fertility, starting growth early in spring and flowering early. The leaves are large and the leaflets always longer than broad (the name broad red refers to the width of the whole leaf, not the individual leaflet). In summer it produces a comparatively small number of erect, usually hollow, leafy stems with rather few nodes (*c.* 5–7), bearing at the top the flower heads, usually light reddish-purple in colour. The buds do not all develop at the same time, so that young, actively-growing shoots are present at the time when the older stems are producing flowers. This results in a rapid recovery if the plant is cut for hay, the young shoots growing rapidly to produce an abundant aftermath or a second hay crop; broad red is therefore often known as *double-cut* clover. The growing-out of a succession of buds during the first harvest year means that few dormant buds are left to continue growth in the following year, and broad red clover is typically a short-lived form, the traditional English varieties lasting for one harvest season only; most new varieties survive however for a second year (this is actually the third year of a ley sown under corn). Its main value is thus for short leys, and it is used, either alone or with ryegrasses as the basis of most one- and two-year leys for silage or where two hay-cuts are taken, or one hay-cut followed by grazing. Annual dry matter yields may be up to some 13 t/ha, with a D-value of about 60, but broad red is essentially a clover for good conditions, and will not grow satisfactorily under conditions of low fertility or on waterlogged or acid soils.

*Cultivars.* It is the usual practice to save broad red clover seed from the second growth, after cutting once for hay, as seed-setting in the first growth may be poor, and the very strong leafy growth makes harvesting difficult. The former practice of harvesting seed from a ley primarily intended for hay is now obsolete, as only certified seed may be marketed. Cultivars available include some local varieties such as Dorset Marl and Drewitts, and various British and continental bred varieties, some of which are autotetraploids, usually more vigorous and higher yielding than the diploids.

Many cultivars are very susceptible to clover rot caused by the fungus *Sclerotinia trifoliorum* and to attack by clover eelworm (*Ditylenchus dipsaci,* an important nematode pest causing a form of clover sickness, and sometimes spread by infected red clover seed). No immune cultivars are available, and although many show some degree of resistance to one or other of the troubles, it is difficult to combine good resistance to both. Thus many high-yielding tetraploids such as Hungarapoly and Red Head show good resistance to clover rot, but are very susceptible to eelworm, whereas the diploid Sabtoron and the older local varieties, although resistant to eelworm, are very susceptible to *Sclerotinia*; the tetraploid Deben, bred at Cambridge, appears to be the most promising cultivar where both troubles are involved. The 1978/9 list includes thirteen diploid and nine tetraploid cultivars.

*Single-cut red clover. (Medium early red clover.)* A type intermediate between broad red and the late-flowering type; somewhat denser and more prostrate than broad red, flowering some two weeks later, and not yielding a second hay-cut. Not now important, since the newer broad red cultivars are higher yielding and almost equally persistent; the 1978/9 national list includes only one cultivar, which is not on the recommended list. The group has, in the past, been rather loosely defined; thus Aberystwyth S 151 has sometimes been listed here and sometimes as a broad red, and the Canadian Altaswede sometimes here and sometime as a late-flowering red.

*Late-flowering red clover. (Late red clover.)* A continental form,

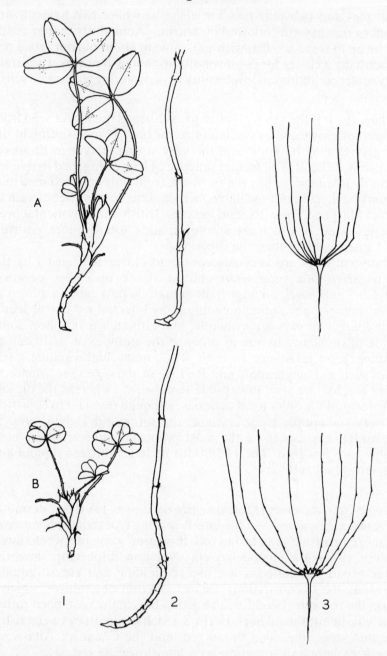

Fig. 43. 1, vegetative shoot in early spring. 2, base of flowering shoot. 3, diagram of habit of flowering plant of A, broad red clover; B, late-flowering red clover.

long cultivated in south-western England and Wales. The true late-flowering reds are longer-lived plants than the broad reds, starting growth later (May) and flowering approximately four weeks later. The leaves are smaller and the leaflets of the leaves produced in spring are almost circular, although those produced later in the year show less difference from broad red. A very dense tuft is formed in spring, and at the flowering period late-flowering red shows many more elongated stems than broad red, each longer, with more nodes (*c.* 7–14), less erect and more nearly solid. Flowers usually a rather deeper red-purple. All stems tend to elongate at the same time, so that there are few or no short shoots in active growth at the flowering period. The plant is thus slow to recover from cutting and produces little aftermath. Numerous dormant buds remain to continue growth the following year, and late-flowering red is a longer-lived type than broad red, often persisting well into the third harvest year and sometimes longer.

Late-flowering red is thus mainly used in leys of two or more years' duration which are to be cut for hay; it gives a rather late but heavy hay-cut, which may exceed that from broad red, but can only be cut once in the year. It is hardy and slightly more tolerant of poor conditions and hard grazing than broad red. Management for seed-production must, of course, be different from that for broad red; no hay-cut can be taken, but the crop may be grazed (or cut early for silage) up to mid-May. This treatment gives a seed harvest (usually in September) at about the same time as that from the second cut of broad red.

*Cultivars.* Two rather similar British local varieties are produced, Montgomery and Cornish Marl. The bred cultivar Aberystwyth S 123 was derived from these two, and in New Zealand the Montgomery variety has been developed into a bred cultivar now known as Grassland Turoa. Some newer cultivars, including the Swedish Merkur and the tetraploid Astra from Aberystwyth (both with fair resistance to clover rot), are available, but the improvements in yield and persistancy of the broad red cultivars have led to some decline in the use of late-flowering red clovers. The 1978/9 Classified List includes twelve cultivars of which four are tetraploids; these are all classified there as medium-late, although the name extra-late-flowering was formerly often used to distinguish this group from the single-cut red clovers.

Red clover seed, taking all varieties together, constituted some 45% of the total usage of herbage legume seed in Britain in the late 1970s.

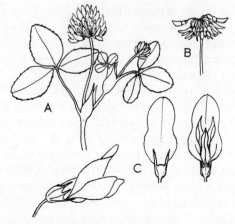

Fig. 44. Alsike. A, flowering shoot. B, flower-head, after pollination. Both × ½. C, side, dorsal and ventral views of single flower, × 2½.

## *Trifolium hybridum* L.   **Alsike Clover**

This is a true species, in spite of its specific name, which may be taken to refer to the fact that it is in some ways intermediate in appearance between red and white clovers. It is not native in Britain, although now fairly well naturalized; its common name is derived from a village in Sweden, from where it was introduced in the early nineteenth century.

*Distinguishing characters.* A short-lived, non-creeping perennial with a growth-habit somewhat similar to that of broad red clover. Leaves glabrous, without leaf-mark, more conspicuously toothed; stipules with long, tapering point, never red-veined. Stems more or less erect, with very short, almost spherical, axillary racemes of white or, more commonly, pale pink flowers. Individual flowers shorter than those of red clover, becoming reflexed after pollination. Pod 1 cm, with usually four to five small seeds (1 mm; 1½ million/kg) which are mainly heartshaped (i.e. with radicle and cotyledon lobes of almost equal size), light or dark green to almost black. Seedling with stalked oval cotyledons *c.* 4 mm long. First leaf *c.* 4 mm long by 5 mm broad, sharply truncate at base, toothed, glabrous.

*Uses.* Alsike is suitable for the same purpose as red clover but, being smaller and lower-yielding, it is of less value. It is, however, rather more resistant to acidity and to waterlogged conditions than red clover and is therefore used as a substitute for the latter where these are likely to cause difficulty. It is a common practice to include a small quantity in a seeds-mixture including red clover as an 'insurance policy', so that there may be some clover present if the red fails to become established. The belief that alsike was resistant to clover rot and stem eelworm was not however borne out in N.I.A.B. trials conducted in the late 1960s.

*Range of types*. Alsike shows comparatively little variation within the species, and its minor importance in Britain would hardly justify the development and maintenance of distinct varieties; several Swedish varieties exist. The greater part of the seed employed is imported from Canada. Under the 1976 fodder plant seeds regulations commerical seed (not certified as to variety) may be marketed and much of the seed sold in Britain is of this commercial type. Two bred cultivars were listed in 1977, the diploid Canadian Aurora and the tetraploid Swedish Tetra. Alsike accounted for some 8% of the total sales of small legume seeds in Britain in the late 1970s.

### *Trifolium incarnatum* L.    Crimson Clover, Trifolium

An annual, introduced from southern Europe, rarely becoming naturalized.

*Distinguishing characters*. Annual, with rather small tap-root. Leaves densely soft-hairy, leaflets very broad, stipules broad, truncate or rounded, often red- or purple-edged. Young plant growing rapidly to form a dense, rosette-like tuft of short shoots, elongating early to form erect stems. Flowers in long, dense terminal racemes; calyx with subulate teeth and stiff hairs, persistent, becoming almost woody in fruit. Corolla bright crimson or white; pod one-seeded, fragile, enclosed in somewhat swollen calyx. Seed large (3 mm, $\frac{1}{4}$ million/kg), yellow to pale orange (white in white-flowered forms), oval, radicle inconspicuous. Seedling large, cotyledons long-stalked, oval, up to 10 mm long. First leaf rounded, *c*. 8 mm, sub-cordate, hairy.

*Use and range of types*. Crimson clover was grown as a winter annual to provide spring keep; its rapid early growth produced plants from an August sowing after corn which were large enough to overwinter successfully. It is not, however, very hardy, and its use was therefore confined to southern England, where it was sown at about 45 kg/ha alone or with Italian rye-grass. The stiff, hairy calyx in the fruiting stage is unpalatable and may be dangerous to stock, and crimson clover must therefore be fed off or cut for hay not later than

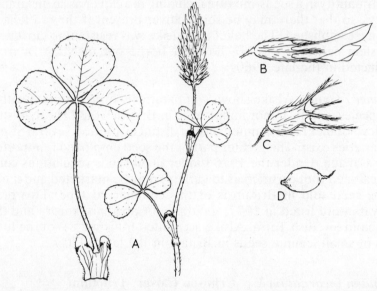

Fig. 45. Crimson clover, *Trifolium incarnatum*. A, single leaf and flowering shoot, × ½. B, C and D, single flower, fruiting calyx, and fruit, × 2.

the early flowering stage (May–early June). It was formerly an important crop for arable sheep, for which it was desirable to provide a succession of keep for as long a period as possible. This cannot be done by successive sowings, since early-sown plants produce elongated stems in autumn and do not over-winter well, while later sowing will give very small plants which again do not survive the winter satisfactorily. A succession of keep could thus only be obtained by the selection of forms with varying dates of maturity, and some half-dozen different types were developed. It is no longer used in Britain, but was included in N.I.A.B. trials in the early 1970s, and may perhaps be re-introduced as a crop in the future.

### *Trifolium repens* L.   **White Clover**

*Distinguishing characters.* A glabrous native perennial, forming small tufts with a tap-root in the early stages; buds in the axils of the rosette leaves soon grow out as horizontal, creeping, leafy stolons, rooting at the nodes. Leaves glabrous; leaflets ovate to nearly circular, conspicuously toothed, usually with pale leaf-mark; stipules small, lanceolate, pointed, connate to form tube around stolon. Flowers in globular umbel-like heads, without bracts, borne singly on long peduncles in the axils of the stolon leaves. Individual flowers on short stalks, erect at first, becoming reflexed after pollination. Corolla white (very rarely pink) forming a short tube; cross-pollinated by hive bees. Pod five- to six-seeded; seeds heart-shaped, bright yellow, darkening to reddish-brown with age, not shiny, small (1 mm, $1\frac{3}{4}$ million/kg with some variation in different forms). Hard seeds frequent, often about 10%. Seedling small, cotyledons stalked, oval, up to 4 mm long. First leaf ovate, slightly toothed, glabrous, lateral veins inconspicuous.

*Trifolium repens* is a tetraploid (4x = 32) and apparently originated either as an autotetraploid from the diploid *T. occidentale* Coombe, an uncommon self-compatible perennial of south-western England and a few other areas of western Europe, or as an allopolyploid from a cross between this and the Mediterranean *T. nigrescens* Viv., a diploid self-incompatible annual.

*Uses.* White clover, typically a long-lived perennial, is essentially a grazing plant; its stems and buds, being all at or near ground level, are not readily eaten off by stock, and grazing therefore removes only leaves unless it is very severe, and hence causes comparatively little damage to the plant. The creeping habit of white clover allows wide vegetative spread, and it can therefore increase rapidly if conditions are favourable. The conditions needed are adequate lime and phosphate and absence of shade; thus if soil conditions are satisfactory, white clover will spread under hard grazing, which prevents its being overshadowed by taller-growing grasses. Such hard grazing is particularly effective in spring, since white clover normally starts active growth later than the better herbage grasses. Under good grassland management, which preserves a satisfactory balance between grasses and white clover, a dense network of stolons is formed between and around the grass shoots, so that nitrogen compounds produced by the clover are readily accessible to the grasses.

White clover is relatively shallow-rooted, and, although not readily killed by drought, makes little growth under dry conditions. Yields vary very much with conditions and with type, and are difficult to assess in the usual mixed grass and clover swards; under good conditions the annual dry matter yield of the larger forms may be up to 5–6 t/ha, and the D-value may be as high as 75.

*Range of types.* White clover shows a very wide range of types, varying almost continuously from the small-leaved wild white type to the very large-leaved Ladino or giant white clover. As leaf size increases there are corresponding increases in petiole length,

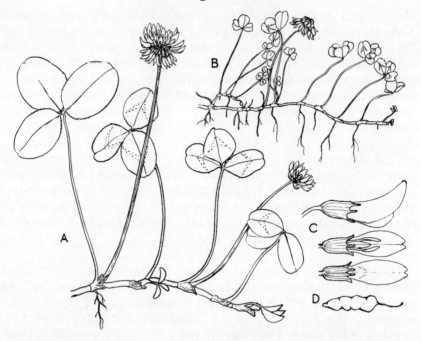

Fig. 46. White clover. A, part of stolon of a large white form (ladino) with leaves and flower-heads. B, of wild white (Kent); both × ½. C, single flower in side, ventral and dorsal views, × 3. D, fruit, × 3.

stoutness of stolons, length of internodes and in the size of inflorescence, flower, fruit and seed. The larger forms usually produce fewer stolons which have fewer adventitious roots and are therefore less tightly rooted to the ground. They are more productive, with a longer growing season; are earlier flowering; better able to withstand competition from grasses under lenient grazing management; are more tolerant of high levels of fertilizer nitrogen; but are less persistent under close grazing and less hardy.

The N.I.A.B. classifies the white clover forms into four groups on leaflet size. The difference in leaflet length between Kent wild white, a typical small-leaved form, and Kersey, a typical medium large form, is arbitrarily set at 15 units (this will usually give a unit of somewhat over 0·5 mm); the groups are then defined as follows:

| | | |
|---|---|---|
| Small | = up to 5 units | |
| Medium small | = 5 to 12·5 units | longer than Kent wild white |
| Medium large | = 12·5 to 20 units | |
| Large | = over 20 units | |

White clover forms also vary in the extent to which they are cyanophoric, that is, capable of producing hydrogen cyanide (prussic acid). Cyanophoric plants contain the glucosides *lotaustralin* and *linamarin,* together with the enzyme *linamarase,* which hydrolizes these substances, with the production of hydrogen cyanide. Only plants with both enzyme and glucoside give cyanide; those with one only, or neither, are acyanophoric. The amount of cyanide produced is usually small (see p. 169), and it does not appear that stock suffer any direct harm from feeding on cyanophoric white clovers. The character is of interest mainly in distinguishing between different white clover forms; for this purpose the *picrate test* is employed. Leaves of the plant to be tested are incubated in closed tubes with strips of filter-paper previously soaked in sodium picrate solution; toluene or chloroform must be added, as no breakdown of the glucosides takes place in uninjured tissue. If the plant is cyanophoric the picrate paper changes from yellow to orange-red.

*Small leaved white clover.* Wild white clover is the wild form, native in Europe, including Britain, and widely naturalized elsewhere. It is a characteristic form of closely-grazed pasture, with small, dark-green leaves, numerous slender branching stolons with short internodes, producing copious adventitious roots at the majority of the nodes, so that stolons are very firmly attached to the ground. Commences growth late, flowering late (end of June), flower-heads small on rather short peduncles. Seed slightly smaller than other forms, but not readily distinguishable. Usually cyanophoric (this character has been shown to be related to winter temperatures; populations from southern and western Europe consist mainly of cyanophoric plants, while those from areas in central and northern Europe with colder winters are mainly acyanophoric).

Wild white clover has been recognized since the pioneer work of Gilchrist at Cockle Park in the early years of the century as a plant of outstanding importance in pasture. Its value lies not so much in its own yield as in its effect on the grasses with which it is growing, although the high protein and mineral content of its foliage make it a valuable food. The nitrogen which becomes available to the grasses via the clover greatly increases their growth, so that although wild white clover grown alone tends to be late-growing and low-yielding, the combination of clover and grass provides one of the best possible swards for long grazing-leys and permanent pasture. Under such conditions the hardiness, permanence and ability to withstand hard grazing which are shown by wild white make it superior to any other leguminous plant.

A large range of wild white forms are thus indigenous or well established in old grassland throughout Britain, but the only listed cultivar is Kent Wild White, produced from selected old pastures. This gives rather low yields (some 2–3 t/ha) but has a very good reputation for persistency. Bred cultivars of similar type include the somewhat higher-yielding Aberystwyth S 184 and the Dutch Pro-nitro, which shows some resistance to clover rot, which can be a serious disease of white clover, particularly in areas where red clover has been frequently grown.

*Medium small leaved white clover.* These are rather larger, more vigorous varieties which are well suited to grazing leys of four or five years' duration. Aberystwyth S 100 is a well known cultivar of this type; Grasslands Huia was developed from white clover naturalized in New Zealand; Sonja is a newer Swedish cultivar with some resistance to clover rot.

*Medium large leaved white clover.* These are the largest forms commonly grown in Britain, higher yielding and better adapted to more intensive management conditions than the smaller forms, but less persistent. They are of high value in medium duration leys, contributing well to silage cuts and to some extent to hay.

The original Dutch White, introduced in the seventeenth century, was a very short-lived form developed under arable conditions, and is now obsolete. Kersey, a local variety developed from plants selected in Suffolk in 1924, has been widely used, but is outyielded by newer Aberystwyth and continental cultivars. Among these are Sabeda, adapted to high rainfall conditions but not resistant to clover rot, and Blanka R.v.P., tolerant of relatively high nitrogen levels, and Milkanova; the two latter both show good resistance to *Sclerotinia*.

*Large leaved white clover.* These are the largest forms, but although high yielding they are intolerant of hard grazing, are not very hardy and tend to be much damaged by slugs in Britain. They are based on the Italian Ladino variety (Giant White) developed under irrigated conditions in Lombardy. Rarely used in Britain but widely grown in the U.S.A.

In the late 1970s white clover seed accounted for 40% of the total usage of herbage legume seed in Britain, and twenty-four different cultivars of this species were included in the Classified List for 1978/9.

## *Trifolium dubium* Sibth.    **Yellow Suckling Clover**

An annual with prostrate stems and small greyish-green leaves. Leaf-lets narrow, terminal leaflet stalked. Flowers pale yellow, in globular axillary heads of from twelve to twenty-five individual flowers, which are at first erect, but become reflexed after pollination, withered corolla persisting. Pod usually one-seeded; seeds ovoid, 1 mm, pinkish-yellow to pale brown, slightly shiny, radicle inconspicuous. Seedling small, cotyledons stalked, oval, 3·5 mm long. First leaf 3 mm long by 4 mm broad, not toothed, glabrous.

A palatable but rather low-yielding annual, common as a wild plant and sometimes sown in short leys, where it may persist by self-seeding. Its main advantage is its ability to establish under dry conditions; it was formerly used mainly in the drier areas of eastern England. It is not included among the crop plants in the United Kingdom national list, but seed, of English origin, 'variety not specified' is occasionally offered for sale.

Seed of yellow suckling clover was formerly a common impurity of

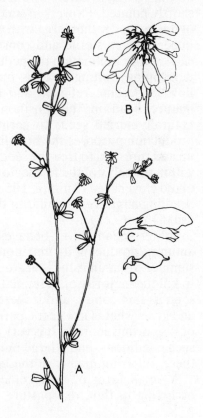

Fig. 47. Yellow suckling clover. A, flowering shoot, × ½. B, flower-head, × 2½. C, single flower, × 3. D, fruit, × 3.

white clover seed but modern methods of seed production have much reduced its incidence as a seed contaminant. Yellow suckling clover is a common weed of lawns, especially on the drier soils.

*Trifolium campestre* Schreb. (*T. procumbens* auct.), Hop clover, is an annual species resembling yellow suckling, but differing in its larger flower heads with from thirty-five to sixty flowers. It is a common, very stemmy, sprawling plant of dry banks, walls, etc., of no agricultural value. *T. micranthum* Viv. (*T. filiforme* auct.), Small Yellow clover, is also similar to yellow suckling, but much smaller, with short-stalked leaves of which the terminal leaflet is sessile, and very small flower-heads of from three to five flowers. A rather common weed of dry lawns.

### *Trifolium subterraneum* L.   **Subterranean Clover** (Sub-clover)

An annual with long, prostrate stems; leaves large, hairy, leaflets very broad, rounded, often with irregular red splashes. Stipules broad, shortly pointed. Flowers in small axillary racemes, the lower flowers of the raceme rather large with white or very pale yellow corollas, the upper flowers sterile and consisting only of a small stiffly-toothed calyx. After pollination the peduncles turn downwards and the whole developing fruiting head is buried in the surface soil; the sterile flowers become reflexed and act as barbs retaining the burr-like mature head in this position. The plant dies off, leaving the naturally-buried seeds to germinate *in situ* in the pod. For seed-production purposes the burr-like fruiting heads are raked or swept up and threshed to free the seeds from the one-seeded pods enclosed within the calyx of the fertile flowers; the seed is almost twice the size of red clover seed (3 mm, *c.* 165 000/kg), dark purple to almost black. Seedling large, very similar to that of *T. incarnatum,* but cotyledons usually more erect.

Subterranean clover behaves as a winter annual, germinating in autumn, producing its maximum growth in spring and dying off in summer. Its main value is in areas where summer drought is sufficient to kill the majority of perennial plants; its buried seeds germinate as soon as rain comes, and it therefore behaves as a self-seeding plant and gives what is in effect a permanent pasture on land only capable of supporting annuals. It is very extensively grown in Australia under such conditions, and a large number of strains have been developed there, differing mainly in the length of growing season required.

A small form of subterranean clover is indigenous in Britain, occurring in thin, dry pasture in the south, but it is of negligible

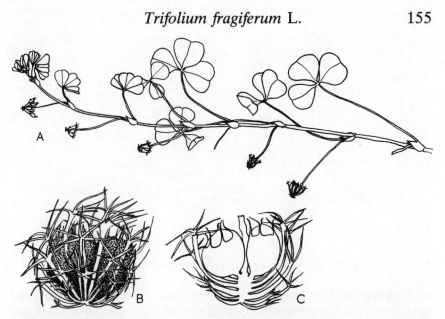

Fig. 48. Subterranean clover. A, shoot with flowers and fruit, × ½. B, mature fruiting head from soil, × 3. C, the same with some sterile calyces removed to show position of fruit.

importance. The much larger Australian varieties grow satisfactorily in southern England, but there appears normally to be little advantage in using them, as the yield does not compare favourably with crimson clover if they are grown purely as winter annuals, or with the larger white clovers if in a longer ley. For effects on animal health see p. 171.

## *Trifolium fragiferum* L.  Strawberry Clover

A stoloniferous perennial resembling white clover, but distinguished in the vegetative stage by its less closely-prostrate habit and its long-pointed stipules, and in the flowering stage by bracts at the base of the individual flower-stalks and by the pale-pink flowers with hairy calyces. In the fruiting stage the plant is very readily recognized; the flowers become reflexed and the calyces much swollen and pink in colour, so that the whole fruiting head has some resemblance to a strawberry. The pods within the swollen calyces are one- to two-seeded; seeds heart-shaped, pale brown with darker flecks, 2 mm, about ¾ million/kg. Seedling very similar to that of white clover, but cotyledons longer, first leaf narrower, more conspicuously toothed.

Strawberry clover occurs occasionally as an indigenous plant in wet pastures in Britain, and is exceptionally tolerant of waterlogging and

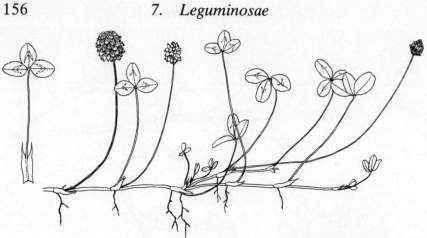

Fig. 49. Strawberry clover. Stolon and detached leaf, × ⅖.

of high salt concentrations. It has been used in the U.S.A., Australia and New Zealand under such conditions where other clovers will not grow, and a number of cultivars, including the large Palestine form, have been developed.

Several other species of clover occur in Britain. Among these may be mentioned *T. medium* L., Zigzag clover, a perennial closely

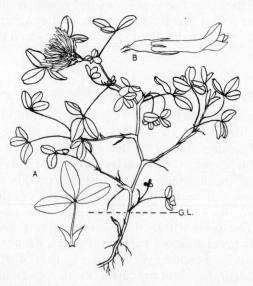

Fig. 50. Zigzag clover. A, flowering shoot and detached leaf, × ½. B, single flower, × 1⅓. G.L, ground-level.

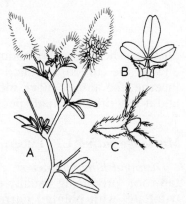

Fig. 51. Haresfoot clover. A, flowering shoot, × ½. B, detached leaf,
× ½. C, single flower after pollination,
× 3½.

resembling red clover in general appearance, but creeping by underground stems and with long, tapering-pointed stipules, and dying to
ground level in winter; and *T. arvense* L., Hare's-foot clover, a small
erect annual of sandy soils, with oblong pink hairy racemes. Clover
species of some agricultural importance in sub-tropical areas are *T.*
*alexandrinum* L., berseem or Egyptian clover, tall growing with
white flowers in round heads; and *T. resupinatum* L., Persian clover, a
winter annual with pink twisted flowers and inflated calyx.

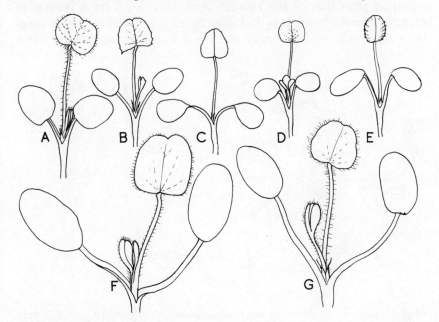

Fig. 52. Clover seedlings, × 2. A, red. B, alsike. C, white. D, suckling. E,
strawberry. F, crimson. G, subterranean.

MEDICAGO

Distinguished from Trifolium by the pointed or mucronate leaflets, and by the curved or coiled pod. Includes the important perennial lucerne, the annual black medick sometimes included in short leys, and a few other annual species of little importance.

## *Medicago sativa* L.   **Lucerne, Alfalfa**

*Distinguishing characters.* A long-lived perennial, with long, stout tap-root, producing usually erect annual stems up to 1 m. Leaves trifoliate, with pointed narrow-ovate toothed leaflets and toothed stipules. Flowers usually blue, in oblong axillary racemes; usually cross-pollinated by bees; 'tripping' necessary; pods smooth, loosely coiled in a spiral. Seeds numerous, *c.* 2 mm, *c.* $\frac{1}{2}$ million/kg, dull greenish-brown, angular, with conspicuous radicle, variable and often distorted and twisted owing to coiling of pod. Seedling with short-stalked, long-oval cotyledons up to 10 mm long. First leaf 6 mm long by 8 mm broad, apex mucronate, base cuneate.

*Growth-habit and use.* The young plant of lucerne is more slender and erect than that of the clovers, and does not form a rosette of leaves at ground-level. It is therefore much less tolerant of grazing at this stage, and can be readily damaged by the removal of buds if

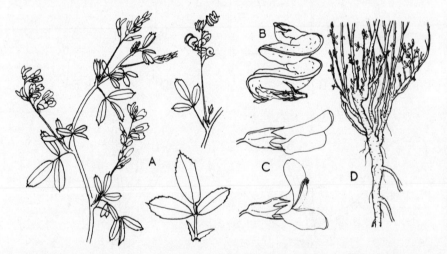

Fig. 53. Lucerne. A, flowering and fruiting shoots and detached leaf, × $\frac{1}{2}$. B, fruit, × 3. C, flowers in early opening and tripped stages, × 2. D, crown and upper part of root of old plant, × $\frac{1}{8}$.

grazed during the first year. In the second year buds at the base of the stem grow out to form erect leafy stems, and these in turn produce further basal shoots so that a branched crown of short perennial stems is formed just above ground-level. When this stage is reached the plant is much more resistant to grazing, but the fact that the bulk of the leaves are borne on the upright stems, well above ground-level, means that photosynthesis, and hence yield and vigour of plant, is very much reduced by continuous grazing. Lucerne can therefore only be used for grazing to a rather limited extent; it is primarily a plant for cutting for feeding green or as hay or silage, or for drying. Used in this way, it is capable of giving three or more cuts per year, and of producing a high total yield. Its deep tap-root makes it very resistant to drought, and it remains productive during a dry summer period when clovers and similar legumes are showing very little growth.

Lucerne is very intolerant of competition, particularly as a young plant, and must either be grown alone or in a specially-designed mixture. In the past lucerne, when grown alone, was sown in rows sufficiently wide apart to allow of inter-row cultivations. The introduction of satisfactory selective herbicides has eliminated this need and lucerne is commonly drilled in rows 10 cm apart at about 13 kg/ha, or sometimes broadcast at about 17 kg/ha. Where lucerne is sown as a constituent of a mixture, a practice much less common now than formerly, the companion species should be sufficiently vigorous to keep down weeds, but not so strongly-growing that they compete seriously with the lucerne. The rye-grasses are too aggressive for the purpose; cocksfoot is satisfactory under dry conditions, but in wetter areas may become dominant, and in such areas timothy or meadow fescue is preferable. Management must be such as to favour the lucerne, and grazing must be controlled, with long rest periods.

The high yield and high nutritive value of lucerne make it perhaps the most valuable of all legumes for stock-feeding, if the agriculture of the whole world is considered. In Britain it is capable of giving annual dry matter yields of over 14 t/ha at a D-value of 60, but since the climate is particularly favourable for the growth of clovers, much more easily managed and only slightly less productive, it is here mainly valued for its drought-resistance, and is more commonly used on the lighter and drier soils of southern and eastern England. Good drainage is in fact always necessary for lucerne, as well as adequate lime and phosphate. Since lucerne is not a frequent crop in Britain, and is not widely naturalized, suitable strains of nodule bacteria are absent from many soils, and seed-inoculation is desirable.

*Origin and range of types. Medicago sativa* is a tetraploid (4x = 32), perhaps derived from the diploid *M. caerulea* Less. ex Led. which is very similar and is indeed sometimes regarded as a subspecies, *M. sativa* subsp. *caerulea* (Less. ex Led.) Schmalh. Both occur, together with other related species or subspecies, as wild plants in the Caspian Sea area. The bringing of lucerne into cultivation and its spread in early times appears to be associated with the spread and increase of importance of the domesticated horse. Lucerne was introduced from Persia into Greece in the fifth century B.C. Its use spread slowly through Europe, reaching England in the seventeenth century. It was introduced into Spain by the Moors, and thence to South America and California, with the Arabic name *alfalfa*; it is now cultivated in almost all temperate and sub-tropical areas. A very large range of types has developed, extending from the wild-type Turkestan forms, which are very hardy but have a short growing season, to the non-hardy types, such as Arabian and Peruvian, which have a very long growing season, but are suited only to warm climates. It is the central part of the range of types which is of interest in temperate climates, and here the adaptability has been increased by crossing with another species. This is *Medicago falcata* L.,* a yellow-flowered hardy species, with small leaves, prostrate stems and sickle-shaped pods, which occurs in Asia and Central Europe, extending into a few areas in eastern England. The hybrids, *M.* × *varia* Martyn (*M.* × *media* Pers.), are very variable and will back-cross with either parent. A whole series of hybrid types is therefore possible, and it is indeed probable that nearly all temperate-climate forms of lucerne now contain some *falcata* 'blood'. The most important hybrid types are the *variegated lucernes,* such as Grimm; the name variegated refers to the flowers, which vary from yellow to blue-purple. Grimm derives its name from that of a Bavarian farmer who introduced a local form of the hybrid lucerne, known as Old Franconian, into the U.S.A. in 1857. The forms previously available in North America had been southern European types, and the introduction of the hardier Grimm enabled lucerne-growing to be extended much further north. Variegated lucernes derived from Grimm are now the main types of the northern U.S.A. and Canada, and they have been reintroduced into Europe.

**Cultivars used in Britain.** Only a small part of the total range of lucerne forms is suitable for use in Britain. Up to about the 1950s

---

* This also is sometimes regarded as only subspecifically distinct, and named as *M. sativa* subsp. *falcata* (L.) Arcangeli, in which case the normal blue-flowered form is *M. sativa* subsp. *sativa*.

some mid-season forms, mainly of the Provence type from southern France and similar types from Hungary and New Zealand were used, together with some late forms, mainly variegated lucerne types of North American origin such as Grimm. These however have too short a growing season to give high yields in Britain.

The cultivars now used are early forms, early in spring growth, early-flowering, with a long growing season extending well into autumn and capable of giving four cuts per year at the early flower-bud stage. Virtually all are blue-flowered forms of the Flamande type from northern France. The older regional French varieties have been replaced by bred cultivars of the same general type. The cultivar Europe, bred in France, is the highest yielding, but is susceptible to wilt caused by *Verticillium albo-atrum*, a disease of increasing importance. Sabilt, from Aberystwyth; Verneuil, from France; and Vertus, from Sweden, although slightly lower-yielding, show resistance to this disease. Plant breeders have made use of other species such as *M. hemicycla* Grossh. from the Caucasus and the North African *M. gaetula* to improve disease resistance, as in the production of Maris Kabul, resistant to *Verticillium* wilt, and Maris Phoenix, resistant to the equally serious but less widespread bacterial wilt caused by *Corynebacterium insidiosum*.

Lucerne seed accounted for some 5% of the total herbage legume seed used in Britain in the late 1970s; eleven cultivars were included in the 1978/9 Classified List, and of these ten were classified as early and one (Everest) as midseason.

## *Medicago lupulina* L.    Trefoil, Black Medick, Yellow Trefoil

*Distinguishing characters.* Annual, or sometimes biennial, with weak, trailing, much-branched stems, leaves trifoliate, leaflets small, ovate, mucronate. Flowers small, yellow, in oval, long-stalked axillary racemes. Pods single-seeded, short, curved, ridged, black when ripe. Seed symmetrical, not twisted, kidney-shaped with conspicuous projecting point near hilum, dull greenish-yellow, 2 mm, about ½ million/kg. Seedling with almost sessile oval cotyledons, 7 mm long. First leaf transversely oval, 4 mm long by 7 mm broad, mucronate, slightly toothed, veins conspicuous, occasional short hairs present. The plant somewhat resembles suckling clover (*Trifolium dubium*), but is readily distinguished by the broader, less grey, mucronate leaflets, the larger, brighter yellow racemes and the very different pods.

*Uses.* Trefoil is a common wild plant of calcareous grassland and can be regarded as useful where it occurs. It was formerly sown in

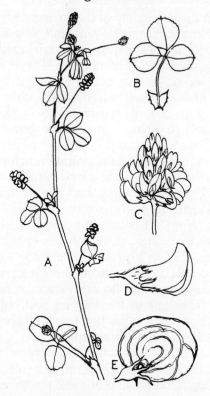

Fig. 54. Black medick. A, flowering shoot, × ½. B, detached leaf, × ½. C, flower-head, × 3½. D, single flower, × 7. E, fruit × 7.

short leys as a poor substitute for red clover and for stubble grazing where its rapid establishment and low seed cost were an advantage. Its use has declined markedly since the middle of the century, and in the late 1970s trefoil seed accounted for only 1% of the total herbage legume use. No special cultivars are listed.

*Medicago arabica* (L.) Huds. (*M. maculata* Sibth.), spotted medick, a hairy, sprawling annual with very broad black-spotted leaflets and large toothed stipules, orange-yellow flowers in short racemes and closely-coiled, almost globular spiny burr fruits, occurs as an unpalatable weed in south-western England, germinating in autumn and dying out in summer, leaving bare patches in a ley.

*Medicago polymorpha* L. (*M. hispida* Gaertn.), hairy medick, with smaller unspotted leaves and flatter burr fruits, occurs occasionally in southern England, but rarely on farmland. Both are cultivated to a

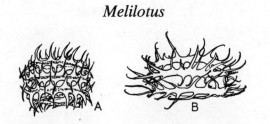

Fig. 55. Fruits of A, *Medicago arabica* and B, *M. polymorpha,* × 3.

small extent as winter cover crops on cotton and maize fields in southern U.S.A. under the name of bur clover. These and other burr-fruited *Medicago* species may become serious weeds of sheep-pasture, as the fruits become entangled in wool and seriously reduce its value.

## MELILOTUS. (SWEET CLOVERS)

Annuals or biennials, with rather woody erect stems, trifoliate pointed leaves and lanceolate entire stipules. Flowers white or yellow, diadelphous, in long axillary racemes, pods one-seeded, straight. Seed yellowish-brown, resembling red clover in size, but with longer radicle. Seedling with almost sessile long-oval cotyledons 9 mm long. First leaf ovate, mucronate, with truncate base, toothed. Epicotyl elongating early.

Melilots are drought-resistant and tolerant of poor conditions, but are not of high agricultural value. If allowed to grow tall, they produce high yields, but the produce is very woody; if cut earlier, before the stems become woody, the yield is low. Growth is erect, with few basal buds, and the plants are therefore not tolerant of hard cutting or grazing. They contain *coumarin* (whence the name sweet clover; cf. sweet vernal grass), which, although sweet-scented, is bitter in taste, and are therefore unpalatable. Spoiled melilot hay or silage may be toxic owing to the conversion of coumarin into compounds closely related to warfarin, used as a rat poison, and which, like warfarin, prevent normal clotting of blood. Melilots are extensively grown in North America under conditions where clovers are unsatisfactory, and where the shortness of the ley prevents the use of lucerne; in Britain they are likely to be of value only for green-manuring, and are rarely sown.

The main species are *M. alba* Medic., White melilot, Bokhara clover, biennial with stout stems up to 1·5 m, flowers white, 4–5 mm long, wings and keel equal. Hubam clover is an annual selection.

*M. officinalis* (L.) Pall., Yellow melilot, Common melilot, is very similar, but with yellow flowers, with wing petals longer than keel. *M.*

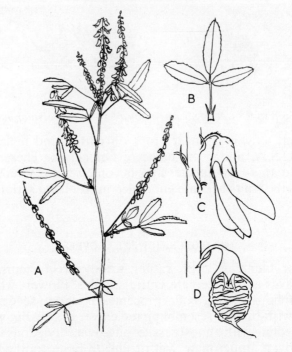

Fig. 56. Yellow melilot (*Melilotus officinalis*). A, upper part of flowering shoot, × ½. B, detached leaf, × ½. C, flower, × 4. D, unripe fruit, × 4.

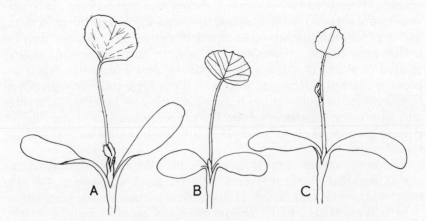

Fig. 57. Seedlings of A, lucerne; B, black medick; C, white melilot, × 2.

*indica* (L.) All., Small-flowered melilot, is a smaller annual with small yellow flowers, occurring occasionally in waste places.

### LOTUS. BIRD'S FOOT TREFOILS

Rather small herbaceous plants with pinnate leaves composed of five leaflets, the lower pair basal and resembling stipules (whence the name trefoil), true stipules minute.

### *Lotus corniculatus* L.   Common Bird's-foot Trefoil

A deep-rooted herbaceous perennial, with stems prostrate or semi-erect. Leaves small; inflorescences axillary long-stalked umbels of from three to six yellow or yellow and orange-red flowers. Calyx teeth erect in bud. Cross-pollination by bees. Fruit a slender cylindrical straight pod 25–50 mm long, with numerous seeds. Ripe pods spreading horizontally so that whole umbel resembles a bird's foot. Seeds shining brown with darker mottling, intermediate in size between red and white clover, *c.* 1·7 mm, 900 000/kg. First leaf of seedling trifoliate.

Bird's-foot trefoil is common in dry pastures on poor land; it is a useful plant on such soils, but is not usually considered worth sowing in Britain. It does not compete well with clovers on the better soils; establishment is slow, and the proportion of hard seeds high (up to 50%). Scottish trials in the late 1960s indicated that bird's-foot trefoil might be a useful alternative to white clover on dry soils of low fertility where grazing pressure was not severe. Cultivated in France and the U.S.A. alone or with grasses, sown at 5–20 kg/ha.

The indigenous British form is a small prostrate plant with broad leaflets. The cultivated forms are larger and more erect; some of these are referred to *L. tenuis* Waldst. and Kit., a diploid with narrower leaflets and smaller flowers, less drought-resistant.

*Lotus uliginosus* Schkuhr. (*L. major* auct.), greater or marsh bird's-foot trefoil, is a larger plant, with larger leaves, wide-spreading stems and short rhizomes. Umbels with eight to twelve smaller flowers, calyx teeth spreading not erect, pods more slender, seeds smaller, 1·2 mm, about 1·75 million/kg.

Occurs in Britain on marshy soils, and is almost the only legume tolerant of such conditions. Rarely sown here, but has been employed in New Zealand for the improvement of marsh land, and Scottish trials in the early 1970s indicated that it has considerable potential for the improvement of wet upland pastures.

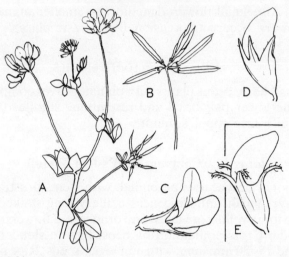

Fig. 58. Birdsfoot trefoil, *Lotus corniculatus*. A, flowering shoot, × ½. B, fruiting head, × ½. C, flower, × 2. D, bud, × 3. E, bud of *L. uliginosus,* × 3.

ANTHYLLIS. KIDNEY VETCH

### *Anthyllis vulneraria* L.   Kidney Vetch

A perennial with stout tap-root and branching crown. Leaves pinnate with from three to five pairs of lanceolate leaflets, terminal leaflet larger; leaves forming a rosette-like tuft in winter. Erect leafy stems in summer up to 50 cm, softly hairy. Inflorescence terminal or sub-terminal, consisting of several short crowded racemes subtended by bracts, each with about thirty almost sessile yellow flowers. Calyx inflated, hairy, with short teeth. Stamens monadelphous. Fruit a one-seeded pod enclosed in persistent calyx. Seed oval, radicle-lobe inconspicuous, 2·5 mm, about 450 000 /kg, upper half green, lower half yellow. First leaf of seedlings simple, obovate.

Kidney vetch occurs occasionally in dry, calcareous pastures. It is rarely sown, but was sometimes employed in eastern England to replace red clover under dry conditions. It gives only a single hay cut, does not compare favourably in yield with red clover, and under ley conditions tends to be short-lived.

ONOBRYCHIS. SAINFOIN

### *Onobrychis viciifolia* Scop. (*O. sativa* Lam.).   Sainfoin

(The name *O. sativa* Lam. is the one used in the E.E.C. classification.)

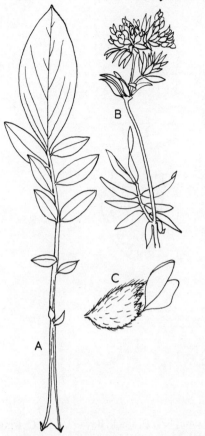

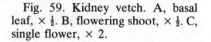

Fig. 59. Kidney vetch. A, basal leaf, × ½. B, flowering shoot, × ½. C, single flower, × 2.

*Distinguishing characters.* A perennial, with stout tap-root, producing numerous erect leafy stems. Leaves pinnate, with numerous pairs of lanceolate lateral leaflets, terminal leaflet small. Stipules broad, pointed, membranous, red in colour, fused to form tube around stem. Long axillary racemes of large flowers. Calyx teeth long narrow, petals pink with red veins; standard semi-erect, wings very small. Stamens diadelphous. Fruit a one-seeded indehiscent pod, with straight ventral suture and semicircular, usually strongly-toothed dorsal suture. Pericarp pale brown when ripe, surface marked with reticulate ridges, about 7 × 5 mm, about 44 000 fruits to kg. True seed, obtained by milling fruits after threshing, large, dark olive-brown, kidney-shaped, about 4 × 3 mm, 66 000/kg. First leaf of seedlings simple, second trifoliate.

*Use.* Sainfoin is almost confined in Britain to the southern half of the country and to chalk and limestone soils, where it provides

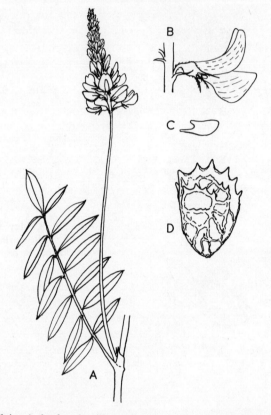

Fig. 60. Sainfoin. A, leaf and axillary raceme, × ½. B, single flower, × 1¾. C, wing petal, × 3. D, ripe fruit ('unmilled seed'), × 3.

valuable sheep-feed and hay. Hay is difficult to make, as the leaflets tend to become brittle before the mid-ribs and stems are dry, but when well-made is regarded as the most valuable of all hays for racehorses. Sainfoin is usually sown under corn, at about 100 kg/ha unmilled 'seed' ('in husk') or 50 kg of the milled seed; 1 kg of unmilled sainfoin contains about two-thirds the number of seeds present in 1 kg of true seed, but the sowing rate is doubled to allow for the greater proportion of hard seeds. Unmilled sainfoin is very likely to be contaminated with the somewhat similar woody fruits of burnet (p. 205), milling of the seed enables this to be cleaned out.

### Types

Sainfoin is cross-pollinated, and a number of local varieties exist; these can be divided into two fairly distinct groups:

*Giant sainfoin,* giving two cuts per year, but persisting only for about two years. Flowers in first year if spring sown, and has long flowering period and comparatively few long stems. Introduced from France about 1830. English Giant is a variety produced in the eastern counties.

*Common sainfoin,* a long-lived perennial persisting for five years or more, but giving only a single hay-cut each year. Does not flower in first year; in second and later years has a shorter flowering period than giant, with more numerous but rather shorter stems. Possibly indigenous, but probably introduced into cultivation from France in seventeenth century. More commonly grown than giant. Cotswold Common and Hampshire Common are well known regional varieties, and the Canadian cultivar Melrose has been placed on the U.K. National List.

The annual dry matter production of sainfoin may be up to 10·5 t/ha; this is lower than that of red clover or lucerne, and yields in the second half of the year may be low. Digestibility is however high, with a mean D-value of 63, and sainfoin may well merit more attention on this account. In the late 1970s sainfoin seed represented less than 1% of the total British herbage legume use. It is much more widely used in Europe and North America, where numerous cultivars exist. The genus is a large one, and some other species may be cultivated.

*Ornithopus sativus* Brot., Serradella, is a Portuguese annual with pinnate leaves, pink flowers and curved, indehiscent pods which break up into one-seeded pieces. It grows to a height of 50 cm, has been used as a fodder and green-manuring crop on poor dry soils in Europe. It is not cultivated in Britain; the much smaller *O. perpusillus* L., Bird's-foot (from the arrangement of the pods; cf. *Lotus*, p. 165), occurs on dry, sandy soils in southern England, but is of no agricultural importance.

### HERBAGE LEGUMES AND ANIMAL HEALTH

Although the herbage legumes provide valuable fodder for livestock they may on occasion have adverse effects on animal health. White clover and some other species may contain cyanogenetic glucosides (p. 151), which although harmless in themselves may be hydrolysed to produce hydrogen cyanide. It is doubtful if their concentration is ever high enough in Britain to put stock at risk, but some cases of poisoning in New Zealand have been attributed to this cause.

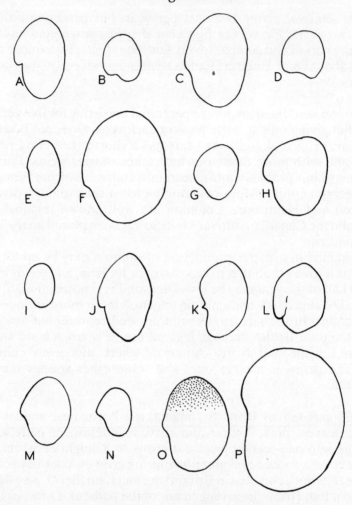

Fig. 61. Outlines of seeds ot clovers and other herbage legumes, × 10. A, red clover. B, alsike. C, crimson clover. D, white clover. E, suckling clover. F, subterranean clover. G, strawberry clover. H, zigzag clover. I, haresfoot clover. J, lucerne. K, black medick. L, white melilot. M, common birdsfoot trefoil. N, greater birdsfoot trefoil. O, kidney vetch (stippling indicates position of green coloration). P, sainfoin (milled).

Herbage legumes may, like members ot the *Cruciferae* (p. 82), contain glucosinolates which can be goitrogenic in effect.

Bloat can be a serious condition, and is most likely to occur when stock are turned out on to lush legume-rich swards after a period of time on a diet containing a high proportion of roughage. Substances causing foaming or inhibiting peristalsis are apparently involved.

Saponins can act as foaming agents, and the use, for example, of low saponin forms of lucerne such as *Medicago hemicycla* and its derivative Maris Phoenix may be of value. Where proteins are implicated it may also be possible to breed for non-bloating characters in some forage legumes.

Cases of infertility in sheep have been linked with the presence of oestrogenic substances in clovers, particularly in *Trifolium subterraneum* and *T. pratense,* and it is inadvisable to put ewes on to clover-rich pastures immediately prior to mating. Apart from failure to conceive there may also be serious problems at parturition.

Several species of clover have been shown to be capable of causing photosensitization in livestock; the mechanism is not fully understood, but outbreaks are most likely to occur at the time of rapid spring growth of the clover.

## LARGE-SEEDED LEGUMES

### LUPINUS. LUPINS

*Lupinus* is a large genus of perennials (including the common garden ornamental *L. polyphyllus* Lindl.) and annuals, of which a number of species are used as fodder and green-manuring plants.

The annual lupins used in agriculture are stout, erect plants up to 1·5 m high, with palmately compound leaves, narrow stipules, and racemes of large flowers with two-lipped calyx and monadelphous stamens; the pods are straight, hairy and contain several large seeds. They are not frost-hardy, but have the advantage that they will grow on light acid soils, where almost all other legumes fail. It is only on soils of this type that they merit consideration; in Britain they have been used to some extent on the lighter soils of eastern England for sheep-folding, silage and green-manuring. Utilization for fodder is complicated by the fact that lupins are often poisonous, owing to the presence of lupinine and other toxic alkaloids. The concentration of the toxic substances is variable, and tends to be higher in the pods and seeds; they are not destroyed by drying, but may be removed by leaching or steaming. Feeding of large amounts of lupins may result in *lupinosis,* the sheep showing jaundice, staggering and loss of appetite; death may result from asphyxia.

Forms of lupins have been developed in Germany, New Zealand and elsewhere which have a low alkaloid content; these are known as *sweet lupins,* and sweet forms of both yellow and blue lupins are available with alkaloid contents of 0·01–0·03% in the seed, as against 0·8–0·9% in the 'bitter' forms. The sweet lupins are to be preferred

Fig. 62. Upper part of flowering plant of A, white lupin; B, yellow lupin; C, blue lupin; all × ⅓.

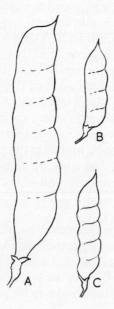

Fig. 63. Pods of A, white; B, yellow; and C, blue lupins, × ½.

for stock-feeding, and can be fed to sheep with greater safety, although care is still necessary.

Lupins were usually drilled at 50–100 kg/ha, either alone or in mixture with rape, oats or sometimes buckwheat. Seed production is difficult, and yellow and white lupins are usually too late in ripening for seed to be produced in Britain. Blue lupins are usually earlier, but loss of seed by shattering (early dehiscence of pods) is a serious problems, and yields are rarely more than 1·25 t/ha. Species which have been or may be occasionally used in Britain are:

### *Lupinus angustifolius* L.    **Blue Lupin**

Up to 1 m high, branching, with rather woody stems. Leaves small with usually from seven to eleven narrow, rather obtuse leaflets, flowers small, pale blue (white forms exist) in short racemes, pods about 5 cm long, with from four to six ovoid seeds. Seeds buff with darker markings, 9 mm, about 4 400/kg. Quick-growing, early ripening.

### *L. luteus* L.    **Yellow Lupin**

Up to 1·1 m high, less branched, leaves larger with from seven to nine broader leaflets; flowers larger, bright yellow, whorled in longer racemes. Pods similar; seeds smaller, 7·5 mm, about 6 600 /kg, white with black markings. Establishment slow under cool conditions, late-ripening.

### *L. albus* L.    **White Lupin**

Up to 1·5 m, stout, little branched, leaves larger with from five to nine ovate-lanceolate leaflets with ciliate margins. Flowers white in continuous racemes, calyx with upper lip entire. Pods large, up to 13 cm, seeds white, flat, large, varying in size up to 15 mm, 1 600/kg. Large-seeded forms have been used as a pulse crop in southern Europe, the seed being apparently safe after steeping and cooking. Used in Britain only as a green-manuring crop, but the introduction of the improved Russian mutant Kievskij in the late 1970s may lead to its cultivation for seed (35–40% protein) in southern England as a possible alternative to beans.

### *L. mutabilis* Sweet    **Pearl Lupin**

Somewhat similar to blue lupin with blue and white flowers turning

Fig. 64. Seedlings of A, white; B, yellow; and C, blue lupins, × ½.

violet and then brown on dying (hence the specific name); seed white or mottled. This species from Andean South America is being investigated as a possible crop. The seeds contain some 46% protein and also 14% oil, and are thus somewhat comparable to soya bean. Forms at present available contain 1·4–2·0% alkaloids but 'sweet' forms could be selected.

### VICIA. VETCHES AND FIELD BEANS

*Vicia* species have weak climbing or straggling stems (stout, erect in *V. faba*), pinnate leaves with small stipules and the terminal leaflet replaced by a tendril or point, flowers in axillary racemes, stamens diadelphous with staminal tube obliquely truncate, pods long, straight, dehiscent. Several weak-stemmed species are common in Britain as wild vetches, the only one commonly cultivated is:

### *Vicia sativa* L.    Common Vetch, Tare

An annual with long straggling four-sided stems and large pinnate leaves ending in a branched tendril, leaflets numerous obovate or oblong, 10–20 mm long, mucronate; stipules lanceolate, pointed, often toothed, with dark mark in centre. Flowers short-stalked, axillary, solitary or in pairs, red-purple or very occasionally white. Pods 50–70 mm, four- to ten-seeded, seeds round, somewhat flattened, varying from pale brown with darker mottling to dark brown, hilum elongated, pale, 4–6 mm, 13 000–18 000/kg. Germination hypogeal, first leaf with two leaflets and short central point.

Grown alone, or more commonly in mixture with cereals, as a forage crop to be folded off or used for silage or occasionally for hay. Valuable for its rapid growth and high yield of green fodder; may be used as a 'smother-crop' to control weeds by its heavy shading effect.

Two main types are used: *winter vetch*, hardy and used for autumn sowings; and *spring vetch*, less hardy, but quicker-growing. The difference is mainly a physiological one; the winter type tends to have smaller seeds and smoother, more cylindrical pods, but these morphological differences are variable and not well-defined.

(*V. villosa*, with flowers in long axillary racemes, is hardier than *V. sativa*, and is known as winter vetch in the U.S.A., where winter forms of common vetch do not survive; it is also commonly cultivated in Central Europe.)

## *Vicia faba* L. (*Faba vulgaris* Moench.). **Field Bean, Broad Bean**

Cultivated in Europe since prehistoric times, but not known wild, and

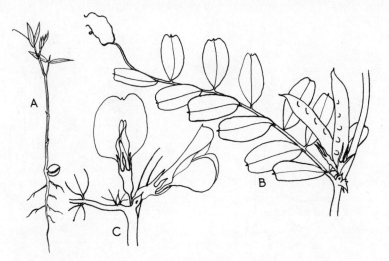

Fig. 65. Common vetch *Vicia sativa*. A, seedling, × ½. B, part of stem with leaf and young pods, × ½. C, pair of flowers in axil of leaf, × 1¼.

of unknown origin. The nearest known wild species is the Mediterranean annual *V. narbonensis* L., but this does not appear to be a direct ancestor.

An erect annual with stout, square, slightly winged stems, large leaves with toothed stipules and usually two pairs of large ovate leaflets; no tendrils are present, but the mid-rib ends in a short, fine point. Short axillary racemes of from two to six large flowers. Corolla white, or occasionally lilac or purple, with black blotch on wing petals. Pods large, straight, fleshy when young and with inner surface downy; seeds varying in size and number in different types. Partially self-pollinated.

Two main types are used within the species: broad beans grown as a vegetable for human consumption, and field beans grown for stock-feeding.

### Broad beans

Seeds large (usually 15–25 mm, 500–1 100 /kg), flat with pale testa, either white or green. Grown in Britain only for picking in the unripe seed stage as a vegetable. A number of fairly distinct cultivars exist, falling into two groups:

(a) *Longpods.* Sufficiently hardy to allow of autumn and very early spring sowing under favourable conditions; pods long, borne several

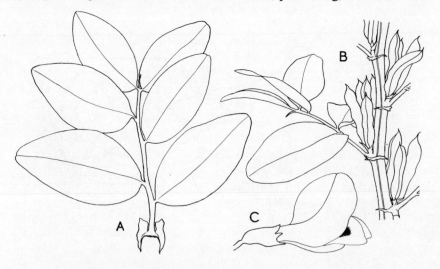

Fig. 66. Field bean (Tick). A, leaf, × ⅓. B, part of stem with leaves and young pods, × ¼. C, flower, × 1⅓.

together and with from four to nine seeds. Aquadulce Claudia is a variety recommended for autumn sowing and Masterpiece Green Longpod one which gives a high yield from spring sowing. Threefold White (flowers, testa and hilum all white) is typical of a number of rather small-seeded cultivars developed for canning and freezing; the testa in these forms is free from tannins and from leuco-anthocyanidins which turn brown on heating, and they are therefore preferred by processors, although of inferior flavour for use fresh.

(b) *Windsors*. Less hardy and suited to spring sowing only; pods shorter, broader and usually borne singly, with from two to five very large seeds. Imperial White Windsor and Imperial Green Windsor are typical cultivars. Less commonly grown than longpods.

Following the E.E.C. classification, broad beans are listed in the National List under the name *'Vicia faba major* L.'.

### Field beans

Seeds smaller (usually 10–17 mm, 1 200–3 500/kg), short-cylindrical or somewhat flattened, testa pale brown darkening with age. Grown on a field scale for harvesting ripe and threshing for stock feed (in prehistoric and early historic times for human food) or alternatively as a constituent of arable silage or forage crops for cutting green. In the 1970s some use has been made of the ripe seeds for processing to produce vegetable protein (cf. soya bean, p. 185).

Cross-pollination occurs to the extent of about 50%, and stocks of field beans tend to be somewhat variable. Some cultivars, such as Throws M.S., are synthetic varieties produced by combining several lines; F.1 cultivar production has proved impracticable owing to the tendency of male sterile lines to lose this character. Cultivars fall into one or other of the following groups:

(a) *Winter beans*. Relatively hardy, slow-growing, stems branching at base. Seed short cylindrical, 1 200–1 750/kg. Cultivars include Throws M.S., Maris Beagle, Bulldog.

(b) *Horse beans*. Suited to spring sowing, less hardy, less branched, quicker growing. Seeds similar to winters, but flatter. Cultivars include Cockfield, Stella Spring.

(c) *Tick beans*. Spring sown, plants similar to horse beans. Seeds smaller, short cylindrical to almost globular, 1 750–3 500/kg. The smaller-seeded cultivars such as Minor are saleable for pigeon food, the larger, e.g. Minden, Maris Bead, Blaze and Dacre, are used for stock-feeding

Winter beans ripen two to four weeks earlier than spring, and can outyield springs by some 600 kg/ha, but have a lower crude protein content (23–24%) than springs (26–28%). They may suffer from winter damage, and then be very susceptible to chocolate spot (*Botrytis fabae*). Springs, if late sown, may be badly damaged by black aphis (*Aphis fabae*); control is by insecticides, but it may be possible to develop genetically resistant cultivars.

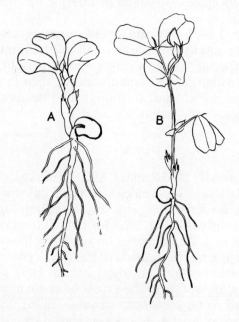

Fig. 67. Seedlings of A, field bean and B, field pea, × ½.

Field beans, with a protein content of round about 25% and some 50% carbohydrates, form a valuable food for stock; digestibility of the cotyledons in all forms is about 88%, but that of the testa in coloured-flowered forms only 10–24% compared with about 60% in the tannin-free white-flowered beans. The bean crop, with the use of residual pre-emergence herbicides, can thus form a valuable break crop in a mainly cereal rotation, but the yield is still somewhat unreliable. Bean 'straw' has a higher protein content than cereal straw and is of potentially high feeding value, but can rarely be harvested in good condition where the seed is allowed to ripen fully.

The National List uses the names '*Vicia faba* subsp. *faba* var. *equina* Pers.' and '*V. faba* var. *minor* (Peterm.) Bull.' for field beans, but does not separate the cultivars into groups.

## *Pisum sativum* L.  **Pea**

A climbing annual, glabrous and glaucous, with slender cylindrical stems from 0·3–1 m, leaves with from two to three pairs ovate leaflets and ending in branched tendril, stipules very large, leaf-like. Flowers from two to four in long-stalked axillary racemes; calyx with broad teeth, corolla white or red-purple, standard broad, erect, stamens diadelphous, stamen tube short transversely truncate. Self-pollinated. Pods smooth, almost cylindrical, with numerous seeds, varying in size (3 000–10 000/kg) and colour in different types. In most forms the inner layer of the pod becomes fibrous and inedible at an early stage. Germination hypogeal, first two leaves of seedling represented by trifid scales of which the shape varies in different varieties.

Cultivated since prehistoric times; probably originating in south-western Asia. Used in the ripe seed stage for human and animal food, and in the 1970s in Canada for the preparation of textured vegetable protein (cf. soya bean, p. 185), in the unripe seed stage as a vegetable for human consumption, or the whole plant as a constituent of green forage or silage mixtures. No truly winter hardy varieties exist, but some hardier forms may be autumn sown in favourable areas.

The fact that peas are normally self-pollinated, and natural crossing very rare, means that a large number of distinct true-breeding cultivars have been developed. These fall into two distinct groups, sometimes treated as separate species, although completely inter-fertile if artificially crossed (for note on nomenclature see below, p. 184); these two groups are the forms used for human consumption, here referred to as garden peas, although now widely grown on a field scale, and those used for animal feeding, referred to as field peas.

### Garden peas (*P. sativum* in the narrow sense)

White flowered with blue-green, white or yellowish seeds, testa colourless, plants tender, stipules unmarked. Cultivated for seeds for human consumption. Cultivars of this type may be arbitrarily divided into two groups according to use:

(a) *Green peas* (*Vining peas*). Grown as a crop for picking green, for use fresh, frozen, quick dehydrated, or canned as 'garden peas'; very variable in height (dependent on internode length) from 0·3 to 1·5 m; taller cultivars as a garden crop only, usually supported on pea-sticks. Length of growing period varies from about ten to fifteen weeks, depending on the number of non-flowering nodes at the base of the stem. Mainly with green, wrinkled seeds, but a few hardier early cultivars have round seeds. Round seeds have a lower water content and harden earlier; they have simple starch grains, while those of wrinkled peas are compound; the sugar to starch ratio at the stage of maturity suitable for freezing is 1:2 for round seeds and 4:1 for wrinkled.

Peas were originally grown as a dry pulse crop for winter use; the immature seeds, green peas, came to be appreciated as an expensive luxury food in the eighteenth century and this led to the development in the nineteenth and early twentieth centuries of green peas as a widely-grown long-picking-period market and private garden crop for consumption fresh during the summer. The development of factory canning and freezing techniques, and the increasing consumer preference for luxury convenience foods, effected a striking change in the middle of the twentieth century, and peas came to be a single-harvest large-scale field crop, fully mechanized and grown under contract to, and very closely controlled by, the processors.

Efficient processing requires a continuous flow of shelled peas to the factory; these must be harvested at exactly the right stage of maturity and processed within a few hours of harvesting. Time of sowing and choice of cultivar must therefore be exactly controlled. Grouping of cultivars on length of growing season is insufficiently accurate, since the actual time to maturity is affected by temperature. A system based on Accumulated Heat Units (AHUs) is employed; AHUs represent the total of the daily differences between daily mean temperature and the minimum (4·4°C) for pea growth. The required total of AHUs from sowing to the particular maturity stage needed is determined in advance for each cultivar.

Maturity stage in the field is measured by mechanical or electromechanical instruments such as the tenderometer, which measures the resistance of the shelled peas to compression and shearing. Tenderometer readings of 95–105 are required for peas to be frozen or quick dehydrated but may be allowed to rise to 115–125 for canning, where slightly more mature peas are acceptable. Harvesting at the freezing stage gives some 75% of the maximum potential fresh weight of crop; harvesting at the canning stage (two days later) some 87%; the maximum fresh weight yield is reached after a further four days (the stage at which peas to be consumed fresh would often be picked); after this the fresh weight yield as well as the quality declines owing to loss of water. Yields are normally in the range of from 4 to 9 tonnes of shelled peas per hectare.

Pods produced at one node all mature at one time, whereas those produced at successive nodes mature in succession; cultivars for use with mechanical single harvest methods are therefore relatively short-stemmed forms of more or less determinate growth pattern with few flowering nodes. Separation of the seeds from the pods by machine shellers makes cultivars with straight blunt-ended pods preferable to those with curved or pointed pods.

The consumer tends to assume that small peas are young peas, and a diameter of about 10 mm is the maximum acceptable; this rules out many of the older cultivars, and even newer ones may reach an unacceptably large size if post-flowering irrigation is employed; on the other hand cultivars with seeds much less than 10 mm in diameter give lower yields. For freezing cultivars with a good green seed are required; this is not essential for canning, where artificial colourants may be added.

With the single-harvest method high plant densities and uniform spacing are necessary for high yields. The optimum economic spacing is of the order of 100 plants per square metre; below about 84/m² yield shows a significant decrease and above 100/m² yield increases do not compensate for the cost of the increased seed-rate. High percentage establishment is essential, and an electro-conductivity test to assess seed vigour is often employed.

Numerous cultivars bred specially to meet the rather exacting demands of the processor and specialist grower are available; these include Sprite, Scout, Puget and Perfected Freezer 70A. Genetic resistance to pea wilt (*Fusarium oxysporum* f. *pisi*) and downy mildew (*Peronospora viciae*) is desirable and is shown by many of the modern cultivars. Frogel is a winter hardy wrinkled-seeded cultivar which may allow harvesting to start earlier.

Some 9–10 t/ha of haulm are left in the field after the mobile viners have finished; this can be made into silage, but its high water content (21% dry matter, with D-value 51 and digestible crude protein 9·5%) means that silage making is not easy.

In addition to this specialized cultivation for processing, green peas are also still widely grown for picking over and marketing in pod for consumption fresh. A few cultivars (e.g. Kelvedon Wonder) can be used in either way, but the majority are distinct; those grown on a field scale for marketing fresh include large-seeded semi-dwarfs such as Onward, as well as true dwarfs. On a garden scale taller cultivars such as Gradus and Alderman may be grown on pea-sticks or other supports. For the earliest crops hardy round-seeded forms such as Meteor and Feltham First are used; these may be sown in autumn or very early spring.

*Sugar peas* (Mangetout) are a distinct form in which the inner fibrous layer of the pod is not developed; the immature pod is cooked whole; not grown on field scale.

(b) *Dry peas.* Cultivars grown on a field scale without supports for harvesting ripe and threshing for human consumption and marketed as dry or packeted peas, or canned after soaking as 'processed peas', or to a smaller extent as split peas after removal of the testa. Grown in the same way as vining peas but at lower plant populations (fifty to seventy-five plants per square metre, depending on type). Yield of threshed seed is 2·5–5 t/ha. Cultivars used as dry peas are of three main types:

(1) *Marrowfats*, with wrinkled seeds and green cotyledons, are considered the best in quality. Mainly fairly large-seeded, *c*. 3 250/kg. The most widely grown cultivars are Maro and Greengolt.

(2) *Blue peas*, with round seeds and green cotyledons; these can be sub-divided into two groups:

*Small blues*, small-seeded cultivars with relatively long straw, suited to the less fertile soils. Seeds usually about 5 000/kg. Vedette is a cultivar being grown to replace the imported American 'Alaska' pea; Frimas is a winter hardy alternative. Used mainly for canning as processed peas.
*Large blues*, large-seeded, short-strawed forms suited to the more fertile soils. Seeds usually about 3 500/kg. Cultivars include Dik Trom, Pauli and Rondo. Used mainly for packeted peas.

(3) *White peas*, with round seeds and yellow cotyledons, mainly cultivars of Dutch or Swedish origin, are grown only for use as split peas. Little grown in Britain, but one leafless cultivar, Filby, was developed in the 1970s.

### Leafless and semi-leafless peas

Breeding work at the John Innes Institute has led to the development of forms with much modified vegetative characters. In semi-leafless peas there are no true leaflets but the stipules remain unaltered; in leafless peas there are no leaflets and only vestigial stipules, the plant relying on the green stem for its photosynthetic area. Cultivars produced in the 1970s show much improved standing ability, better light penetration through the canopy, better air circulation and therefore less likelihood of fungal damage, fewer pods in contact with the soil and easier harvesting of both vining and dried peas. Such cultivars may make it possible to grow peas satisfactorily in areas previously considered unsuitable because of high rainfall.

### Field peas (*P. sativum* var. *arvense* (L.) Poir.)

Sometimes treated as a separate species *P. arvense* L. Flowers bicoloured, purple and red, with seeds brownish or mottled owing to coloration of testa, plants hardier than those of garden peas, stipules with purple coloration at base. Grown only for stock-feed, for ripe seed or cut green for forage or silage, either alone or in mixture with cereals. (It may be remarked that, while these are commonly known as field peas, the field acreage devoted to them in Britain is considerably exceeded by that of the garden peas.) Field peas may be grouped into:

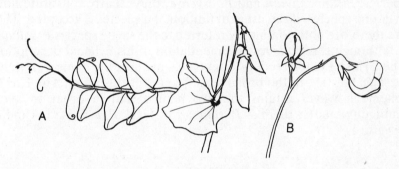

Fig. 68. Field pea (Dun). A, part of stem, with leaf and young pods, × ⅕. B, flowers, × ½.

(1) *Dun peas*, with large seeds, testa dull brown, cotyledons yellow, early ripening, medium-length 'straw'.

(2) *Maple peas* (also called Partridge), with smaller seeds, testa brown, speckled, cotyledons yellow, usually later than duns, and with longer straw. Used, like duns, for stock-feeding, but also has special sale for pigeon-feeding, for which purpose small peas with light hilum are preferred. Named bred varieties are Marathon, small-seeded, storing well, hilum black, and the earlier, even-ripening Minerva with light hilum, higher-yielding and shorter-strawed than Marathon.

(3) *Grey peas*, seeds smaller, spherical, testa pale grey speckled violet, cotyledons yellow. Late ripening. Little grown in Britain, used in southern U.S.A. as winter cover crop (Austrian winter peas).

Peas provide useful stock-food of approximately the same composition as beans, but, as with beans, the yields are very variable. Drilled
at about 40–200 kg/ha according to size and use, in 18–36 cm rows;
suited to medium or light soils. Harvesting difficult except under dry
conditions; in wetter areas usually grown in mixture with cereals to
provide support. They may also, like field beans, be included in
arable silage and forage mixtures. Special forage cultivars, described
as *fodder peas*, were introduced from Germany in the mid-1970s;
these, when harvested as green haulm, give dry matter yields of about
7·5 t/ha, with about 25% crude protein.

### Note on nomenclature of peas

Garden and field peas were originally named by Linnaeus as separate
species, *Pisum sativum* and *P. arvense*; they clearly constitute however one species, if the usual definition of a species is accepted. They
are therefore both commonly referred to the single species *P. sativum*
L. although this involves an emendation of his original description.
The British Plant Breeders' Rights scheme uses the name *P. sativum*
L. sens. lat. (i.e. in the broad sense) to cover this point. The E.E.C.
scheme however retains the two names, and in order to avoid
ambiguity, quotes them as *P. sativum* L. (excl. *P. arvense* L.) and *P.
arvense* L.

GLYCINE

### *Glycine max* (L.) Merr. Soya Bean

Erect annual, 0·6–1 m; leaves trifoliate with large ovate, mucronate,
hairy leaflets; narrow pointed stipules present at base of leaflets as
well as of leaves. Flowers small, yellowish-white, in short axillary
racemes. Pods straight, hairy, with from three to five seeds. Seeds
varying in colour (cream, yellow, green, brown or black) and in size
(5–10 mm, 5 000–10 000/kg) in different forms. Eastern Asiatic in
origin, derived from the wild form *G. soja* Sieb. et Zuss.; cultivated in
China for at least 4 000 years. Very numerous cultivars exist.

Soya bean seeds are of exceptionally high feeding value, with some
30–50% protein, 13–25% oil and 14–24% carbohydrates (figures
based on air-dry seeds with 5·0–9·5% moisture). No starch is present.
Yellow-seeded forms are richer in oil, black-seeded richer in protein.

*Uses*. Soya beans were developed in China as a crop for human
consumption in a wide variety of ways; unripe seeds used as a veget-

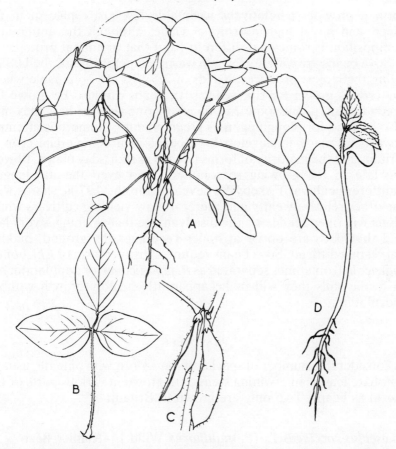

Fig. 69. Soya bean. A, whole plant in fruiting stage, × ⅕. B, leaf, × ⅕. C, pods, × ½. D, seedling, × ½.

able, ripe seeds ground for soya flour or fermented for soy sauce or germinated for bean sprouts. Extensively developed in the U.S.A. during the twentieth century to become the major source of vegetable oil on a world scale. The edible semi-drying oil is used for margarine manufacture, as a salad and cooking oil, and has numerous industrial and pharmaceutical uses; it contains some 59% linoleic acid.

Soya beans, either whole or after oil extraction, are extensively used in the preparation of animal feeding-stuffs and of soya flour for human consumption. Extracted soya meal is the most widely available source of relatively concentrated vegetable protein, and is extensively used as a replacement for, or addition to, animal protein in human diets. For this purpose it is processed as 'textured vegetable protein' to give it something of the texture of muscle fibre; in this form it provides a relatively acceptable low-cost replacement for meat, and has a high nutritional value, although the amino-acid composition is somewhat different from that of animal proteins.

Soya beans are widely grown in warm temperate areas, the U.S.A. being the largest producer. Yields of 2·5 t/ha are obtainable where the crop is grown for ripe seed; soya beans can also be grown for green fodder, giving some 25 t/ha green, or 5–10 t/ha hay; they may also be used for ploughing in as green manure. Numerous attempts have been made to develop soya beans as a commercial crop in Britain, but the majority of forms behave as short-day plants, flowering late in the growing season here, and even the day-length-indifferent cultivar Fiskeby V, developed in the 1970s, shares with the other relatively early and therefore low-yielding cultivars which alone will ripen seed here, the disadvantage that ripening is very late and that pods are borne at nodes very near the ground, making harvesting difficult. Soya beans require a distinct form of *Rhizobium radicicola* (sometimes separated as *R. japonicum*) for nodulation, but in fertile soils they will make apparently healthy growth without inoculation.

### PHASEOLUS

A considerable number of species of *Phaseolus*, with pinnate, usually trifoliate leaves and twining stems, are grown in various parts of the world as beans. Two only are grown in Britain:

### *Phaseolus coccineus* L. (*P. multiflorus* Willd.).    **Runner Bean**

Tall perennial with fleshy root, grown as an annual. Inflorescences axillary, longer than leaves, with from twenty to thirty flowers, scarlet, white or bicolour, pods broad, rough, green; seeds large (10–25 mm, 600–1 200/kg.), usually pink with black mottling, or white in white-flowered forms. Mainly self-pollinated. Germination hypogeal. Grown only for immature pods used as a vegetable. Of Central or South American origin.

*Cultivars*. Most cultivars are climbing forms to 2 m or more high, grown either on supports or as 'pinched' crops in which the terminal buds are periodically removed to keep the plants bushy. Frequent picking over is necessary to ensure continued flowering and hence a succession of pods over several weeks. Typical red-flowered cultivars are Prizewinner, with lightly speckled seeds, and Princeps, with seeds more heavily black speckled. Painted Lady has bicoloured flowers

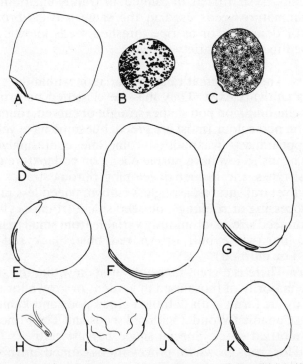

Fig. 70. Seeds of various large-seeded *Leguminosae,* × 2. A, white lupin. B, yellow lupin. C, blue lupin. D, common vetch. E, tick bean. F, broad bean. G, Scotch horse bean. H, blue pea. I, marrowfat pea. J, maple pea. K, soya bean.

and light and dark brown speckled seeds; Sunset has pink flowers and seeds of the Princeps type. White-flowered, white-seeded cultivars such as Emergo and Desiree are preferred for canning or freezing. Some naturally dwarf bushy forms have been developed, which can be grown without supports or pinching; Hammonds Dwarf Scarlet and Hammonds Dwarf White are cultivars of this type. No forms suitable for mechanized single harvesting exist.

**Phaseolus vulgaris** L.    **French Bean** (Kidney Bean; known in U.S.A. as Field Bean)

Annual, of Central or South American origin, climbing or dwarf. Inflorescence with fewer flowers than in *P. multiflorus*; pods narrower, smooth, of various colours. Self-pollinated. Seeds smaller (5–20 mm, 1 800–4 600/kg) but varying much in size and colour in different cultivars. Germination epigeal. Grown for utilization in three ways: as immature pods used as a vegetable in the same way as runner beans, fresh, frozen, or canned; or (rarely in Britain) as older but still immature seeds used in the same way as broad beans, flageolet or demi-sec; or as ripe threshed seeds known as haricot beans, used in a wide variety of ways.

*Cultivars.* There is a great range of variation within the species and numerous cultivars exist. They may be classified on growth habit (dwarf or climbing); on pod shape (straight or curved, round or flat in section); on pod colour (most are green, but some have yellow pods, waxy in appearance—'waxpods', or sometimes confusingly described as 'butter beans', a few have purple pods); on pod texture ('string' or 'stringless' types, the former developing fibrous strands along the dorsal and ventral sutures; stringless cultivars need less preparation before processing or cooking); on seed shape (round, cylindrical or kidney); on seed size (continuously variable from small to large); and on seed colour (white, buff, brown, red, blue, black; self-coloured, bicoloured or mottled).

*Cultivars.* There is a great range of variation within the species and types; the majority of those used in gardens or grown for marketing fresh are dwarf forms with flattened green pods and kidney-shaped seeds, e.g. Canadian Wonder with red seeds and The Prince with red and buff mottled seeds. Special single harvest cultivars have been developed for crops grown for processing as canned or frozen French beans. These show a reduction in branching; this is accentuated by growing them at close spacing (28–42 plants/m$^2$, in rows 40 cm apart; 20 cm rows give rather higher yields but cause difficulties in mechanical harvesting). They are mostly of the stringless, straight, round-section 'pencil-podded' type, but the flavour and texture of this type is considered inferior to that of the flat-podded forms, and attempts are being made to produce cultivars of the latter type suitable for machine handling. Cultivars used in the 1970s include the white-seeded Cascade and Harvester, and Tendergreen with purple and buff mottled seeds.

Cultivars intended for the production of dry ripe seed are grown to only a very limited extent in Britain and are mainly used in warmer drier areas. Most are white-seeded, and some of the larger-seeded forms of these are imported and marketed in Britain as haricot beans. Some coloured-seeded cultivars may also be used in this way, e.g. Royal Red Kidney, Brown Dutch, Mexican Black. Navy beans are particular dwarf forms with small, rounded, white seeds which are imported into Britain on a large scale, mainly from North America, and processed as baked beans. Attempts made in the 1970s to develop cultivars of this type which will meet the rather stringent requirements of the processors, and which can be grown in the more favourable south-eastern areas of England, have met with some success; cultivars used are Purley King and Revenge. Limelight, a cultivar with larger, flatter seeds has been tried in an attempt to find a substitute for the imported butter bean (*Phaseolus lunatus*, see below).

Other *Phaseolus* species not grown in Britain but commonly cultivated in warmer areas and sometimes imported include:

*P. lunatus* L., Lima bean, butter bean, of Central American origin and now widespread. Plants resembling *P. vulgaris*, climbing or dwarf; flowers cream, pods short, flattened, with two to four larger (500–1 000/kg) usually white seeds. Smaller-seeded forms (1 cm) used mainly for immature seeds, in same way as broad bean; those with larger flatter seeds (3 cm) for ripe dry seeds, which are imported into Britain as butter beans.

*P. aureus* Roxb., green or golden gram, mung bean, is probably of Indian origin and widely cultivated there and elsewhere in tropics. Bushy annual with clustered yellow flowers; pods slender, cylindrical, with up to fifteen small (25 000–33 000/kg), rounded, olive green seeds. Used as green beans, or ripe seed as pulse or germinated in dark as Chinese bean sprouts, a useful vitamin source.

*P. mungo* L., black gram, urd, resembles *P. aureus* in area of origin and use and in appearance, but has paler flowers in smaller clusters, pods more erect, hairy, with up to ten seeds of similar size or slightly larger; used in similar ways and as forage and green manure.

*P. angularis* (Willd.) Wight., adzuki bean, probably originated in Japan and is an important pulse crop there and elsewhere in eastern Asia. Flowers yellow in short axillary clusters; pods cylindrical, usually straw-coloured; seeds oblong, maroon, straw, brown or black in colour with long white hilum, 5 000–10 000/kg.

### OTHER GENERA

Crop plants belonging to other genera of the *Leguminosae* which are cultivated in warmer parts of the world, and the products of which may be imported into Britain include:

*Vigna unguiculata* (L.) Walp., cow pea (this is an aggregate species including *V. sinensis* (L.) Savi ex Hassk. and *V. sesquipedalis* (L.)

Fruw.), is an annual of African origin widely cultivated in warm areas. Resembles *Phaseolus* species, with climbing and dwarf forms, inflorescence few-flowered, flowers white, yellow or violet, keel curved (not coiled as in *Phaseolus*). Pod short to very long, with eight to twenty seeds from 2 to 12 mm long, 4 000–10 000/kg, very variable in colour. Typical cow peas (black-eyed pea, black-eyed bean), with white testa and black ring surrounding the white hilum, are used as a dry pulse crop and also eaten as immature pods or immature seeds; these and other forms are also used as fodder, hay and green manure crops.

*Lens culinaris* Medik. (*L. esculenta* Moench.), lentil, is a very ancient Mediterranean crop plant. A slender vetch-like annual with pinnate, tendril-bearing leaves, axillary inflorescences of one to four small white or pale blue or pink flowers. Pods short, flat with one or two lens-shaped seeds (the word lens being taken from the Latin name of the plant), grey to light red speckled with black, 3–9 mm, averaging about 50 000/kg. Widely used as split lentils, i.e. the usually orange-yellow cotyledons only, but also used for lentil flour.

*Cicer arietinum* L., chick pea, is a long cultivated much-branched annual, probably originating in western Asia. Somewhat vetch-like, but without tendrils, glandular-hairy; flowers axillary, solitary, white, green, pink or blue; pods short, swollen, one to two seeded; seeds of characteristic shape, angular with distinct beak, 5–10 mm in diameter, 3 700–6 000/kg, white or variously coloured. Grown in Mediterranean and cooler tropics; will tolerate poor conditions. The most important pulse crop in India; ripe seeds used whole or as flour, green pods eaten as vegetable.

*Arachis hypogaea* L., ground nut, peanut, monkey nut, is an important crop plant of South American origin now widely cultivated in tropical and subtropical countries. A low-growing, tufted annual with pinnate leaves and yellowish flowers turning down after pollination, the gynophore (base of ovary) elongating to bury the developing fruit in the ground. Ripe pods fibrous, containing two to three pea-like seeds, red or brown, globular or elongated according to variety. The seeds contain 38–50% oil and some 30% protein. Ground nut is, after soya, the second most important source of vegetable oil on a world scale. The edible non-drying oil, with some 27% linoleic acid, is widely used for cooking, as a salad oil, for margarine, and in soaps and lubricants; the protein residues are widely used in animal feeding and to some extent in human foods. Considerable quantities of nuts are eaten raw, roasted and salted, in confectionery and as peanut butter.

*Trigonella foenum-graecum* L., fenugreek. Erect Mediterranean annual with long, pointed, narrow pods containing yellow, flattened, very angular, strongly-scented seeds, sometimes ground and the meal included in cattle feeds as spice; an ingredient of curry powder and used pharmaceutically.

*Ceratonia siliqua* L., carob bean, locust bean, is a large tree of the Mediterranean area, belonging to the distinct sub-family *Caesalpinioideae*. Pods large, broad, flattened, fleshy, with high sugar content; formerly imported whole for stock-feeding, now commonly processed for extraction of gums (used in fabric printing) from seeds and residual meal used as ingredient in feeding-stuffs.

**Key to leguminous temperate crop plant genera: vegetative characters**

Leaves trifoliate.
    Leaflets large, 4 cm wide or more.
        Leaves glabrous, stem twining .................*Phaseolus*
        Leaves hairy, stem not twining ...................*Glycine*
    Leaflets small
        Leaflets emarginate or obtuse ...................*Trifolium*
        Leaflets mucronate.
            Stipules lanceolate or broad, toothed ..........*Medicago*
            Stipules linear, not toothed ...................*Melilotus*
Leaves with five or more leaflets.
    Leaflets palmately arranged .......................*Lupinus*
    Leaflets pinnately arranged.
        Terminal leaflet present.
            Leaflets 5, lower pair stipule-like ...............*Lotus*
            Leaflets numerous.
                Terminal leaflet largest ...................*Anthyllis*
                All leaflets similar.
                    Stipules membranous, broad. ............*Onobrychis*
                    Stipules green, minute .................*Ornithopus*
        Terminal leaflet replaced by tendril or point.
            Stipules smaller than leaflets ......................*Vicia*
            Stipules larger than leaflets ......................*Pisum*

# 8

# OTHER CROP-PLANT FAMILIES

The dicotyledonous families which include the agricultural crops of major importance in Britain have been dealt with in separate chapters. This chapter is concerned with the other families which include crop plants which deserve mention, but which are of less importance in British agriculture. The families are treated in botanical order, starting with those regarded as more primitive. The crops concerned are of very varying type: some are agricultural crops occupying only a small fraction of the total cultivated area; these include some crops formerly widely grown, but now obsolescent here, and some which have been grown experimentally, but which have not so far become established in Britain. Others are primarily horticultural rather than agricultural crops. A number are mainly crops of warmer climates, mentioned here because they are occasionally grown in Britain, or because their products are commonly imported.

### PAPAVERACEAE

*General importance.* Contains only one agricultural crop plant, oil poppy, grown in various temperate areas for its oil-containing seeds. Grown only experimentally in Britain. The plant is also the source of the drug opium.

*Botanical characters.* A small family of mainly herbaceous plants. Leaves alternate, exstipulate, often pinnately divided. Flowers usually large, solitary, actinomorphic, insect-pollinated. Sepals two, free; petals four, free, crumpled in bud. Stamens numerous, free; gynaecium superior, of two to many united carpels, unilocular, with numerous ovules on parietal placentas. Style very short or absent, stigma capitate or in the form of sessile rays on top of ovary. Fruit a capsule, often dehiscing by pores. Seed small, often reniform, with sculptured testa; endospermic, with very small embryo. The family includes a number of common weeds.

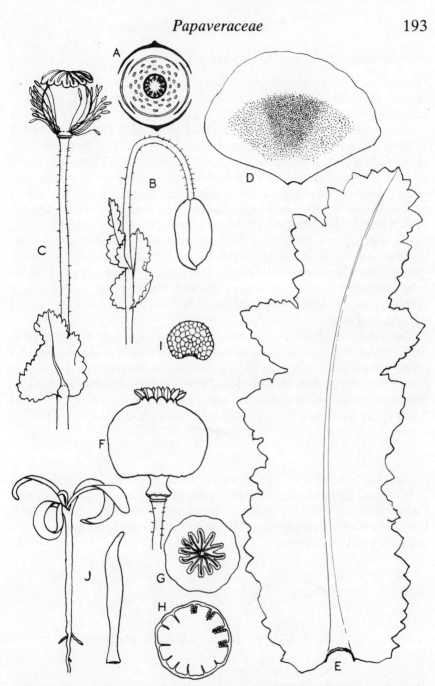

Fig. 71. Oil poppy, *Papaver somniferum*. A, floral diagram. B, bud. C, flower with petals fallen. D, detached petal. E, basal leaf. F, G, H, ripe capsule in side and top view and transverse section. All × ½. I, seed, × 10. J, seedling and detached cotyledon, × 3.

### *Papaver somniferum* L.   **Oil Poppy**

An annual, 1–1·5 m high, glaucous with scattered stiff hairs. Leaves sessile, clasping, oblong, up to 30–10 cm, shallowly lobed, coarsely crenate-serrate. Stem stout, little branched. Flowers large, nodding in bud, erect in flower and fruit, on stout peduncles. Sepals caducous (falling when flower opens); petals large, 10 cm broad by 7 cm long, crumpled in bud, white with basal red-purple patch. Ovary globular, of nine to fourteen united carpels, top convex, marked by stigmatic rays; placentas forming incomplete vertical partitions not extending to centre, and bearing very numerous ovules. Fruit an ovoid or almost globular capsule, 5–6 cm in diameter, glabrous and glaucous, dehiscence mechanism present in form of ring of pores below the flat, stellate, oblong-toothed stigmatic cap, but not functional in cultivated varieties. Seed reniform, 1–1·5 mm, *c.* 1·6 million/kg, testa pale grey-blue or almost white, marked with raised reticulations. Germination epigeal, expanded cotyledons linear, sessile, about 12 mm × 2 mm.

The seeds contain a valuable edible drying oil with some 65% linoleic and 20% oleic acid; seed yields of about 1 t/ha are obtainable, giving an oil yield of some 350 kg. Seeds can also be eaten whole in confectionery. A fine seed-bed and very shallow drilling are necessary owing to the small size of the seed; a close stand with plants about 5 cm apart in 45 cm rows is required for high yields.

### LINACEAE

*General importance.* The *Linaceae* includes only one species of agricultural importance. This is *Linum usitatissimum*, grown in two forms: one, linseed, for the oily seed; the other, flax, for its stem fibres, from which linen is manufactured. Neither form is now commonly grown in Britain.

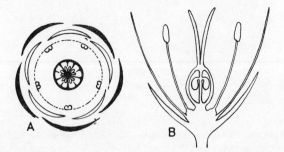

Fig. 72. *Linaceae.* A, floral diagram. B, vertical section of flower.

*Botanical characters.* A small family, mainly herbaceous, with alternate simple leaves. Inflorescence usually cymose, flowers conspicuous, insect-pollinated. Sepals and petals five, free. Stamens five, sometimes with five staminodes. Gynaecium of five carpels, joined but with free styles; fruit usually a septicidal capsule. Placentation axile, ovules anatropous. Seeds with scanty endosperm, cotyledons with oil and protein food-reserves.

### *Linum usitatissimum* L.   **Linseed** and **Flax**

#### *Description*

An annual, with erect, wiry stems, bearing numerous lanceolate entire leaves, glabrous and greyish-green, 3–4 cm long. Stems (under closely spaced conditions of field crop) little branched, except near apex, where a series of short branches bear terminal flowers 2–2·5 cm in diameter. The sepals are small and lanceolate, the petals wedge-shaped, crumpled in bud, bright blue, or in some forms white. Stamens five, with filaments broadened at base. Ovary ovoid, with five erect styles, and originally five-chambered, with two ovules in each loculus, but becoming almost completely divided into ten chambers by *false septa* growing in from the mid-rib of each carpel, and almost meeting the placenta between the ovules. Nectar is secreted by the disc at the base of the filaments, but under temperate conditions the plant is almost entirely self-pollinated. The ovary enlarges after fertilization to form a spherical or somewhat flattened capsule. In the majority of cultivated varieties the capsule is almost

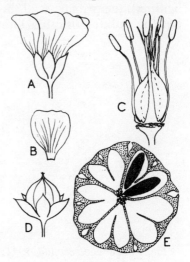

Fig. 73. Linseed. A, flower. B, petal, × 1½. C, flower, with petals and sepals removed, × 4. D, young fruit surrounded by calyx, × 1½. E, diagram of transverse section of ripe capsule, × 4 (two seeds only shown).

completely indehiscent, and the seeds are set free only on threshing, or by irregular shattering.

The seeds are oval, flattened, 4–6 mm (90 000–200 000/kg), pale to dark brown, distinctly shiny. The outer epidermis of the testa consists of large cells, which readily absorb water and swell up to form a structureless mucilage. The seeds contain 35–40 % oil and about 20% protein, and form a very valuable animal food. They can be fed complete, or as linseed cake after extraction of the bulk of the oil. The extracted linseed oil is a valuable drying oil, used in the manufacture of paints, linoleum, etc., and also as human food. Care must be taken in the feeding of unextracted linseed owing to the presence of a cyanogenetic glucoside. If the seed is allowed to soak in cold water, enzyme action may result in the breakdown of this glucoside, with the production of the poisonous compound hydrogen cyanide (prussic acid); boiling water should therefore be used in the preparation of linseed gruel for calves, etc.

Germination is epigeal; the cotyledons elongate and become blunt-lanceolate in shape. The stem elongates rapidly and bears alternately arranged leaves, the lowest very close to the cotyledons.

### Stem anatomy: fibres

The stem shows a typical dicotyledonous structure, with a ring of vascular bundles surrounding the central pith. Immediately outside the phloem of each bundle is a group of fibres; the individual fibres are very long, slender, thick-walled cells, 25–50 mm long by about 20 $\mu$m thick. They differ from many fibre cells in being only very slightly lignified; the primary wall shows some lignification as the plant ages, but the bulk of the thickening consists of cellulose. The extracted

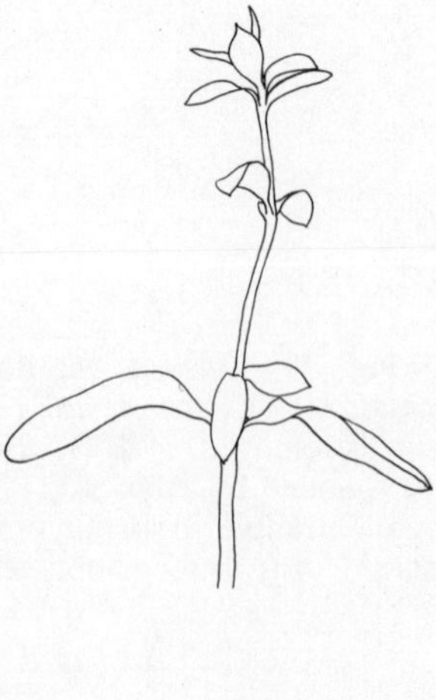

Fig. 74. Linseed seedling, × 1½.

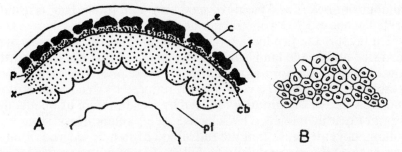

Fig. 75. Flax. A, diagram of part of a transverse section of a mature stem. B, a single small fibre-group more highly magnified. *e,* epidermis. *c,* cortex *f,* fibres. *p,* phloem. *cb,* cambium. *x,* xylem. *pi,* pith.

fibres are therefore soft and flexible, but extremely strong, and it is to these characters of the fibre cells that flax owes its importance as a textile fibre. The fibres are extracted by *retting*, which is a process of controlled rotting, formerly carried out by immersing the pulled plants in ponds, or exposing them to damp air (dew-retting), but now completed more quickly and precisely at controlled temperatures in flax factories. The retting results in the breakdown of the middle lamella of the cell walls, so that the cells can be readily separated. Mechanical treatment (scutching and hackling) of the retted stems removes the xylem and other tissues, leaving the linen fibre.

### Origin and range of types

*Linum usitatissimum* appears to have originated in south-west Asia, and has been in cultivation for some 5 000 years. Archaeological evidence suggests that the closely related *L. bienne* Mill. (*L. angustifolium* Huds.) native in the Mediterranean area and western Europe, including England, was also used. This species may therefore have played some part in the development of the cultivated forms. As the crop spread, specialization took place, with the Mediterranean and European forms developing as fibre plants—that is, flax—and the quicker-maturing forms grown in the warmer climates of southern Asia and particularly India, giving rise to typical linseeds, grown purely for seed.

There have thus developed two distinct cultivated forms within the species; both are largely self-pollinated and have therefore remained distinct, although crosses between the two are possible.

*Flax.* Grown for fibre-production, and typically a plant of cool, moist climates. Stems tall, little branched except in the upper part, number of flowers and yields of seed low, seed small. Flax is usually drilled at about 190 kg of seed per hectare, so as to give a dense stand which favours the production of slender, unbranched stems. A good seed-bed and fertile land is necessary, and, as flax is not frost-hardy, sowing should not normally take place before mid-April. Clean land is essential, as the yield is considerably reduced by weed competition, and the value of the produce reduced by an admixture of weeds; flax is rather readily damaged by selective weed-killers. Harvesting is by pulling, not cutting, so that the full length of stem is secured; it takes place soon after the petals have fallen. The immature capsules are combed off before the straw is retted; the seed yield under these conditions is, of course, low—perhaps 350 kg/ha.

During the early part of the twentieth century the flax grown in the British Isles (mainly in Northern Ireland) was largely derived from seed of north European origin, but from about 1930 improved varieties were bred in Northern Ireland. Increased yield and higher quality were first obtained by the development of taller-growing forms, but progress along these lines could not be continued very far, since tall plants are likely to lodge. The quality of the fibre from a lodged crop may be impaired, and the harvesting costs are greatly increased if mechanical pulling becomes impossible. Further progress therefore depended on the production of cultivars with a higher percentage of fibre in the stem, giving a higher yield of scutched flax from the same weight of crop; for quality to be maintained this increase had to be due to an increase in the number of fibres, and not to a greater diameter of individual fibre cells. These objectives were largely achieved, and the introduction of such cultivars as Stormont Gossamer and Liral Prince resulted in the raising of the yield of scutched flax from under 500 to about 800 kg/ha. Even with this improvement, competition from imported fibre, and particularly from alternative synthetic fibres, proved too strong, and flax was unable to maintain its place as an important economic crop in Britain.

*Linseed.* Grown for the valuable oil and protein content of the seed; typically a crop for warm summer conditions. Plants are usually shorter, quicker-maturing and considerably more branched than flax, the capsules more numerous and the seed usually (but not always) larger. High seed yields cannot be combined with good quality fibre, and the 'straw' of linseed is a by-product of little value, occasionally used for the extraction of coarse fibre for packing, etc.

The area devoted to linseed cultivation in Britain has never been large, and most of the linseed used has been imported. The high

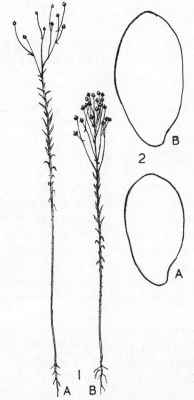

Fig. 76. 1, Mature plant, ⅕; 2, seed, × 7, of A, flax and B, linseed.

nutritional value of the seed sometimes made it a worthwhile crop, and small areas were occasionally grown for consumption on the farm. Larger scale cultivation was encouraged in the 1940s to offset war-time shortage, and the area reached some 35 000 ha in 1948, but fell again rapidly as imported supplies became available. Interest in linseed as a cash crop revived in the 1970s with the appreciation of its potential value as a break crop in a predominantly cereal rotation, and with the availability of an E.E.C. subsidy. Seed is usually drilled at about 90 kg/ha; this is a lower rate than for flax, and gives a more open stand allowing greater branching of the individual plants. Linuron or other selective herbicides may be used, and the crop is combined when the seed is fully ripe; care is necessary on account of the wiry stems and very free-flowing seed. Expected yields are in the range from 1·5 to 2·5 t/ha.

The small area of linseed in Britain does not justify the breeding of special varieties for use here, and the cultivars employed are of North American and north European origin.

## VITACEAE

*General importance.* Includes only one crop plant, the grape vine, widely grown in warm-temperate climates, and to a small extent in favourable areas in England.

*Botanical characters.* A small family of some 500 species, mainly woody climbers with tendrils. Leaves alternate, flowers numerous, inconspicuous, greenish, pentamerous with ovary of two joined carpels, superior. Ovules basal, few; fruit a fleshy berry.

*Vitis vinifera* L. Grape vine. Woody perennial with deciduous palmately lobed leaves; tendrils and inflorescences arising opposite leaves. Fruit (grapes) produced in large dense panicles; ripe fruit with high sugar content. Epicarp (skin) adherent to pulp, white or dark purple. Numerous cultivars, which are clones, propagated by grafting; some American cultivars are derived from crosses with other species. Some American species are also used as root-stocks since they are resistant to the root aphis *Phylloxera vastatrix*, to which *V. vinifera* is very susceptible.

Probably originating in south-western Asia, but long cultivated in Europe; wild forms mainly dioecious, cultivated hermaphrodite. Mainly grown for wine production in specialized vineyards; support, pruning, frequent spraying and hand harvesting needed. High quality wines from special areas, dry, usually calcareous slopes, often terraced, giving low total fruit yield, but high sugar content. Some cultivars used for production of dessert grapes; raisins, sultanas and currants are dried grapes; some parthenocarpic (seedless) forms are used.

## ROSACEAE

*General importance.* A family of outstanding horticultural importance, containing many of the main cultivated fruits, but with very few members of agricultural value. Forage burnet is an agricultural crop of very minor importance; hawthorn, and to a lesser extent blackthorn, are farm hedging plants.

*Botanical characters.* A very large family, including numerous shrubs and small trees. Leaves alternate, usually stipulate, often compound. Inflorescence racemose; flowers actinomorphic, insect-pollinated, very variable in structure. Calyx of five free sepals, sometimes with an epicalyx; petals five, free; stamens usually ten or more;

gynaecium of from one to many carpels, sometimes united, with one or more ovules, the style of free carpels often lateral or basal. The flower is usually perigynous, but in some forms the carpels are so deeply sunk in the receptacle that the flower becomes epigynous, while in others the central part of the receptacle is strongly convex, bearing the carpels on its outer surface. The fruiting receptacle may be either dry or succulent, as may also the pericarp, so that a very wide range of fruit types is found.

The members of the *Rosaceae* grown as fruit crops are almost all vegetatively propagated, and the horticultural varieties are clones. Methods of propagation vary; in strawberries a natural means of vegetative spread is present in the runners, while in raspberries and similar plants the shoots which arise from below ground-level, and which naturally produce adventitious roots, can be separated and replanted. In the tree fruits, however, no natural means of vegetative propagation are present, and cuttings rarely produce adventitious roots. For these plants it is therefore necessary to resort to budding or grafting, the clone which it is desired to propagate being used as the *scion*. The *root-stock* is a form of the same or a closely-related species which can either be readily grown from seed, or of which rooted cuttings can be obtained, usually by means of the *stool-bed* method. This involves the cutting-back and earthing-up of stock plants in such a way as to encourage the development of rooted shoots from below ground-level; these can be later separated and replanted as stocks. The type of root-stock used has no genetic effect on the scion, but does markedly affect the size, vigour and longevity of the resultant tree; it is therefore an advantage to use root-stocks vegetatively propagated by the stool-bed method, since these are clones of uniform type, whereas 'free' stocks grown from seed will usually show considerable variability. The East Malling series of root-stocks for apples, e.g. Malling IX, are examples of such clonal stocks.

Although seeds are not used in propagating fruit crops (except in the production of new varieties), it is essential that seed should be set in order that full development of the fruit may take place. Many clones either produce little or no viable pollen, or are partially or entirely self-incompatible, that is to say, will only set seed when cross-pollinated. Plantings of one single variety are useless in such cases, and interplanting with a suitable pollinator is necessary. This must be another variety which is compatible with the first (cross-incompatibility is common e.g. in cherries), which produces sufficient viable pollen, and which flowers at the same time. An adequate number of pollinating insects, usually bees, is also necessary.

The family can be divided into a number of sub-families. These include the *Rosoideae*, in which the carpels are separate and borne on a persistent receptacle which may be either convex or deeply concave. In *Potentilla*, which includes a number of weed species but no crop plants, the convex receptacle is dry and the carpels develop as achenes. In *Fragaria*, the strawberry, achenes are also formed, but the receptacle becomes enlarged, succulent and edible. The modern cultivated strawberry varieties are clonal forms of the octoploid *Fragaria* × *ananassa* Duch., a series of hybrids between the two wild octoploid species *F. virginiana* Duch. from North America. and *F. chiloensis* Duch. from South America. They are large-fruited, with glabrous leaves. Usually self-compatible, but some varieties, e.g. Tardive de Leopold, are male-sterile and must be interplanted. Other species are the diploid *F. vesca* L. (wild strawberry, cultivated forms of which are known as Alpine strawberries) with hairy leaves and small fruits, and the tetraploid *F. moschata* Duch. (hautbois, of European origin, now almost obsolete as a crop), also with hairy leaves, and with larger purplish fruit with achenes on the upper part of the receptacle only.

*Rubus* has the receptacle dry, but the carpels develop as small drupes (drupels or drupelets), the mesocarp forming the succulent edible part, while the endocarp, the inner layer of the ovary wall, forms the hard 'pip' containing the single seed. The genus includes a considerable number of species and hybrids forming a polyploid series. *Rubus idaeus* L., the raspberry, is a diploid with pinnate leaves and succulent drupelets separating from the receptacle when picked; extensively grown commercially; fruiting on second-year canes (or current year's in autumn-fruiting cultivars), mechanical harvesting of fruit to be processed may be possible with suitable cultivars. *R. fruticosus* L. is a convenient aggregate name for the very variable blackberry group, which has palmate leaves, spiny stems, and drupelets which separate less readily from the receptacle. The group is usually divided into numerous species, with chromosome numbers varying from diploid to hexaploid. Normal sexual reproduction, apomixis, and vegetative reproduction by tip-rooting of long trailing stems are all possible, and hence very numerous forms exist. Many are wild and are common weeds of much-neglected grassland; a few are cultivated, including some spineless forms. The hybrid *R.* × *loganobaccus* Bailey, the loganberry, is a hexaploid derived from a cross between a raspberry and a tetraploid American blackberry; the leaves are pinnate and the elongated dark-red aggregate fruit resembles the blackberry in the drupelets not separating readily from the receptacle. Numerous other Rubus forms of hybrid origin exist

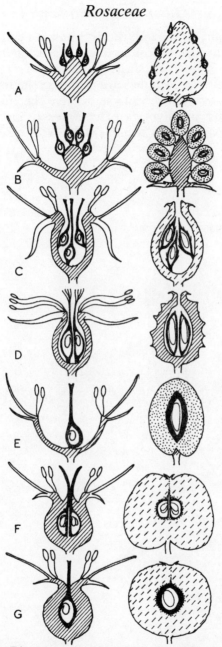

Fig. 77. *Rosaceae*. Diagrammatic vertical sections of different types of flower and fruit. A, strawberry. B, blackberry. C, rose. D, burnet. E, plum. F, apple. G, hawthorn. The receptacle is marked by diagonal shading; broken diagonal lines in the fruiting stage indicate that the receptacle is succulent. Carpels solid black; parts of carpel-wall which are succulent in fruit stippled. Not to scale.

and are occasionally cultivated; the odd-number polyploids are usually of lower fertility and therefore lower yield than the even-number
forms.

In the genus *Rosa* the achenes are borne on the inner surface of a
deeply concave succulent receptacle, to give the characteristic rose
hip. Many species and hybrids of complex origin are cultivated,
vegetatively propagated, as ornamentals; wild roses, including *R.*

Fig. 78. Forage burnet, upper part
of flowering shoot, $\times \frac{1}{2}$.

*canina* L., dog rose, often apomictic, are common hedgerow plants. Some hips are good vitamin C sources.

*Sanguisorba minor* Scop. subsp. *muricata* Briq. (*Poterium polygamum* W. & K.), forage burnet, belongs to a genus in which the achenes are reduced to one or two, borne inside the small very deeply concave dry receptacle. It is a tufted perennial growing to a height of about 60 cm. The radical leaves are pinnate, with numerous toothed ovate leaflets; the rather wiry,erect, flowering stems bear numerous leaves of similar structure, but with rather narrower leaflets. Ovate toothed stipules are present at the base of the leaves. The plants are monoecious, or sometimes polygamous, with the usually unisexual flowers crowded together in short, ovoid racemes. The lower flowers of the raceme are male, and the upper ones female; hermaphrodite flowers may be present in the middle. Four green or purplish sepals are borne on the rim of the tubular receptacle; petals are absent. The male flowers have numerous stamens, the female flowers usually two one-seeded carpels sunk in the receptacle and each bearing a long style with red, brush-like stigma; pollination is by wind. The receptacle persists in the female and hermaphrodite flowers, and tightly encloses the mature achenes. The 'seed' thus consists of achenes and hardened receptacle; it is pale brown, about 6 mm long, oval in side-view, roughly rectangular in cross-section, with four toothed, longitudinal wings and with the surface between the wings coarsely toothed and reticulated.

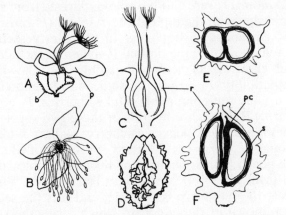

Fig. 79. Forage burnet. A, female flower, × 4. B, male flower, × 2. C, vertical section of female flower to show position of carpels. D, ripe fruit, × 4. E, transverse and longitudinal sections of fruit, × 6. *b*, bract. *p*, perianth. *pc*, pericarp. *r*, receptacle. *s*, seed.

Forage burnet is an introduced plant, of Mediterranean origin, sometimes grown as a grassland 'herb'. It is deep-rooted, drought-resistant and relatively high-yielding, but may become somewhat unpalatable if allowed to develop its rather hard flowering stems. It does not persist well in herb mixtures on non-calcareous soils. It has become widely naturalized on calcareous soils, and is now perhaps more important as a weed than as an intentionally sown crop. The diaspore is sometimes found as an impurity of unmilled sainfoin 'seed', which it somewhat resembles in size and appearance.

*S. minor* Scop. subsp. *minor* (*Poterium sanguisorba* L.), salad burnet, and *S. officinalis* L., greater burnet, are related plants which may occur as unsown constituents of some grassland, the former on dry calcareous soils and the latter in wet areas.

*Prunoideae*, the plum sub-family. Trees and shrubs with simple leaves; receptacle concave, not persistent; one carpel, fruit a single drupe. *Prunus* is a large genus, divisible into several sections or subgenera. In the plum section (leaves deciduous, rolled in bud) *Prunus spinosa* L., blackthorn or sloe, is a spiny shrub, with white flowers produced in spring before the small lanceolate leaves. The fruit is a small drupe with purple skin and green flesh. Blackthorn is a common shrub in farm hedges, but less satisfactory than hawthorn or quick (see *Crataegus*, below) and rarely planted. The cultivated plums, *P. domestica* L., are allo-hexaploids derived from a crossing of the tetraploid sloe with the diploid cherry-plum (*P. cerasifera* Ehrl.) followed by chromosome doubling. In the cherry section (leaves deciduous, folded in bud) the diploid *P. avium* L. includes the cultivated sweet cherries, all self-incompatible and showing marked cross-incompatibility. *P. cerasus* L., sour and Morello cherries, is a tetraploid, usually self-compatible. Duke cherries, *P.* × *gondouinii* (Poit. et Turp.) Rehder, are hybrids of these two species.

In the almond section (leaves as cherry, axillary buds in threes, endocarp rough) *P. persica* (L.) Batsch., probably of Chinese origin, includes the peach (epicarp hairy) and nectarine (epicarp glabrous). In *P. dulcis* (Mill.) Webb (*P. amygdalus* Batsch.), the almond, from western Asia, the mesocarp is inedible, and it is the seed which is eaten after removal of the endocarp which forms the shell. In the laurel section (leaves evergreen) *P. laurocerasus* L., cherry-laurel, is a shrub or spreading tree with oval, leathery leaves, small white flowers in erect leafless racemes and black, cherry-like fruits. It is commonly planted in garden hedges and is often naturalized; it is an undesirable shrub on farms, as it is markedly *poisonous* owing to the presence of a cyanogenetic glucoside in the leaves.

*Maloideae*, the apple sub-family. Flowers epigynous, with from one to five carpels sunk into the receptacle and fused with it. Fruit a pome. Haploid chromosomes 17.

*Pyrus communis* L., common pear, has five carpels with styles free throughout their length. Each carpel contains two ovules, and the endocarp becomes horny, forming the core. Stone-cells are present in the 'flesh' of the enlarged receptacle. Cultivated varieties often self-incompatible; diploid, triploid and tetraploid varieties exist. Fruits mainly on short spur-shoots.

*Malus*, the apple genus, has fruits of similar structure, but with the styles joined at the base, and stone cells absent. *Malus domestica* Borkh. (*Pyrus malus* L.), the cultivated apple, is of hybrid origin, probably derived from several wild species. Cultivars are mainly diploid, some triploid; self-incompatibility is frequent; distinguishable into dessert, culinary and cider apples. Traditional widely spaced large trees are tending to be replaced by closely planted bushes or other forms in which vegetative growth is strictly controlled; choice of root-stock is important.

*Cydonia oblonga* Mill., the quince, has a fruit similar in structure to that of the apple, but with several ovules in each carpel; grown occasionally for fruit and some forms also used as a root-stock for pears.

*Crataegus* has the fruit a pome, but the endocarp of the carpel wall becomes hard, so that the seed is enclosed in a 'stone'.

*C. monogyna* Jacq., hawthorn, whitethorn or quickthorn, is a spiny shrub or small tree, with deeply-lobed leaves and conspicuous stipules. The small white flowers, produced in corymbs in May and June, have only one carpel, single seeded; the pome is red, about 10 mm in diameter. Hawthorn is the best and commonest plant for farm-hedges; non-poisonous, spiny, hardy and withstands laying. Propagated by seed, which requires a period of after-ripening and germinates slowly.

*C. laevigata* (Poir.) DC. (*C. oxyacanthoides* Thuill.), Midland hawthorn, is a very similar shrub, with more shallowly-lobed leaves and two carpels; less commonly grown.

*Mespilus germanica* L., the medlar, has a rather hard, somewhat apple-like open pome, with the tops of the five stony carpels exposed. Edible only after breakdown of receptacle tissue has begun (bletting); now rarely grown.

### GROSSULARIACEAE

*General importance.* A small family including black and red currants and gooseberries, grown for their fruit.

*Botanical characters.* Deciduous shrubs with palmately lobed alternate leaves, flowers usually pentamerous with inferior ovary of two joined carpels, ovules numerous, placentation parietal, fruit a berry.

*Ribes nigrum* L. Black currant. Leaves glandular, with pointed serrate lobes. Vegetative buds large, giving rise to long erect first-year shoots, on which axillary racemes of greenish flowers are produced in the second year. Berry black with characteristic flavour and high vitamin C content. Cultivars are clones, readily propagated by cuttings and grown mainly for jam and syrup. Mechanical harvesting possible by cutting whole stems. but this entails cropping in alternate years only. Use of a mobile stripper avoids this difficulty.

*Ribes rubrum* L. (*R. sativum* Syme) Red and white currants. Leaf lobes blunter, buds smaller, racemes borne on older wood, not suited to harvesting of whole stems; less important commercially.

*Ribes uva-crispa* L. (*R. grossularia* L.) Gooseberry. Spiny shrubs with smaller stiffer deeply lobed leaves; flowers borne singly or in pairs. Fruit green, yellow or red when ripe, epicarp hairy in many cultivars. Ripe fruit for dessert; more commonly harvested green for cooking.

### CUCURBITACEAE

*General importance.* The *Cucurbitaceae* is a family of horticultural importance, the main crops being cucumbers, largely grown as a glasshouse crop, and marrows grown as a non-hardy outdoor crop.

*Botanical characters.* A mainly sub-tropical family, with most members large soft-stemmed annuals, sprawling or climbing by tendrils. Vascular bundles irregularly arranged with phloem on both sides of xylem (*bicollateral*). Separate male and female flowers; calyx and corolla both of five joined segments. Stamens in male flowers five but with two pairs joined and so appearing as three. Gynaecium in female flowers of three joined carpels, inferior; placentation axile, but placentae often filling the three loculi and giving the appearance of parietal placentation. Fruit a characteristic partly succulent berry known as a *pepo*.

Species of two genera are commonly grown in Britain; these are *Cucurbita* (marrows and pumpkins), with shallow-lobed corolla, stamens all joined by their anthers, and tendrils branched, and *Cucumis* (cucumbers and melons), with deeply divided corolla, free anthers, and unbranched tendrils. One species of a third genus *Citrullus* (water melon) may very occasionally be grown under glass; in this the flowers are similar to those of *Cucumis*, but the tendrils are branched and the leaves pinnately, not palmately, lobed.

### *Cucurbita pepo* L. **Marrow**

*Cucurbita pepo*, with pointed-lobed stiffly hairy leaves, and the fruiting peduncle angular, includes the majority of the forms known as vegetable marrows in Britain and as squashes in the United States. Plants trailing, with long prostrate stems, or bush, with short stems; not frost hardy. Fruit is very variable in shape; cylindrical or somewhat angular in the forms commonly grown in Britain; green, white or striped. The fruit is at first soft and edible throughout, and may be harvested at this stage as courgettes. For this purpose special cultivars, some of which are F.1 hybrids, are usually employed; these are early fruiting, with short erect stems, bearing up to twenty fruits per plant harvested when 10–20 cm long, yielding perhaps 35 t/ha during the season. The receptacle and epicarp later harden; marrows are usually harvested as unripe fruits some 25–50 cm long, giving a total yield of, say, 70 t/ha. Fully ripe fruit is occasionally used for jam; fruits not suitable for long storage. Forms with small inedible hard fruits of various shapes and colours are sometimes grown as ornamental gourds.

C. *maxima* Duch., with cylindrical fruiting peduncle, includes the pumpkin, with very large flattened-globular fruits, and some long-keeping winter squashes (e.g. Hubbard squash) which are suitable for winter storage, but little grown in Britain. C. *moschata* Duch., with peduncle widely expanded, and C. *mixta* Pang. with hard corky peduncle, are softly hairy species with leafy calyx lobes, cultivated in warm-temperate and sub-tropical areas. C. *ficifolia* Bouche, a perennial with hard globular striped fruits and black seeds, has been used as a resistant root-stock on which to graft young cucumber plants. All these *Cucurbita* species are of Central American origin.

### *Cucumis sativus* L.   **Cucumber**

The cucumber, which probably originated in India, has roughly hairy leaves, acute corolla lobes, and the fruit with either black or white spines in the early stages. Immature fruit used; extensively cultivated under glass to provide cucumbers throughout the year. English glass-house types have long cylindrical fruits, dark green and smooth at harvest, parthenocarpically produced. Many cultivars are F.1 hybrids, some bearing female flowers only. Ridge cucumbers are hardier forms with shorter fruits, not developing in the absence of pollination, grown outdoors. Gherkins are similar forms in which the very young fruit is harvested for pickling. Numerous other forms exist, including ones with globular and yellow or white fruits, but these are not grown commercially.

*Cucumis melo* L., melon, has softly hairy leaves, obtuse corolla lobes, and the fruit without spines. Fruit sweet when ripe, often scented. Numerous forms exist, including netted melons with reticulate and Cantaloupes with longitudinally furrowed fruits. Some are grown under glass in Britain, but most melons used are imported.

*Citrullus lanatus* (Thunb.) Mansfeld (*C. vulgaris* Schrad., *Colocynthus citrullus* (L.) O. Ktze.), water melon, is widely cultivated in the tropics, but is of little importance in Britain. The fruit has a high water content, with the placentae large and succulent, so that the seeds are embedded in the edible parenchyma, and not confined to the centre as in melons.

POLYGONACEAE

*General importance.* The *Polygonaceae* is a family in which the weed species are of much greater importance than the crop-plants. The only agricultural crop is buckwheat; rhubarb is the most important of the very few horticultural crops. Family name derived from the genus *Polygonum,* which includes a number of common weeds.

*Botanical characters.* A small family of mainly herbaceous plants, with alternate simple leaves. Stipules united to form a membranous tube surrounding the stem, known as the *ochrea.* Inflorescences racemes or axillary clusters or whorls; flowers small, actinomorphic, hermaphrodite or unisexual. Perianth variable, either of six segments, in two often dissimilar whorls of three; or five segments, spirally arranged; petalloid (and then usually pink or white) or sepal-like. Stamens variable in number, often six or nine. Gynaecium superior, of three (rarely two) united carpels, with a single basal orthotropous ovule. Pollination by wind or insects. Fruit three-cornered, with hard leathery pericarp; it is strictly a nut, since it is formed from three joined carpels, but is usually referred to as an achene. The enclosed seed has a curved embryo embedded in copious mealy endosperm.

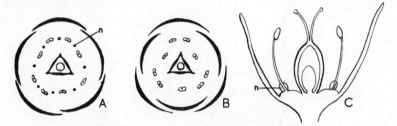

Fig. 80. *Polygonaceae*. A, floral diagram of *Fagopyrum*. B, of *Rheum*. C, vertical section of flower of *Fagopyrum*. *n*, nectary.

### *Fagopyrum esculentum* Moench.    (*Polygonum fagopyrum* L.).   Buckwheat, Brank

Buckwheat is an annual with rather weak erect stems, swollen at the nodes. The leaves are alternately arranged, cordate or triangular in shape with sagittate base. The lower ones are large on long petioles, the upper smaller and almost sessile. The inflorescences are stalked axillary clusters; the individual flowers are cymosely arranged and each cluster contains flowers of very varying age. Flowers consist of (usually) five white or pink petaloid perianth segments, eight stamens, and three united carpels forming the triangular superior ovary, bearing three short styles. Small spherical nectaries are present between the bases of the stamens, and pollination is by insects, mainly bees. The flowers are dimorphic—that is, two types of plant are found, one with long filaments and short styles, the other with flowers in which the anthers are on short filaments and the styles longer. Crossing normally takes place from a long-styled to a short-styled plant, and vice versa. The perianth segments do not increase in size after pollination, but the ovary enlarges to form a three-cornered achene, which is the agricultural 'seed'. The ripe achene is (in the form usually grown) greyish-brown in colour, slightly shiny, acute-angled, about 6 mm long, about 40 000 /kg. The pericarp and the testa (husk) form about 40% by weight of the whole achene, the remainder being white, floury endosperm (starch-parenchyma) with the slender, curved embryo embedded in it.

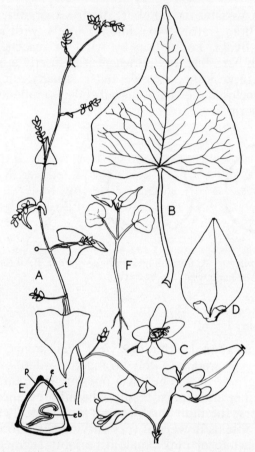

Fig. 81. Buckwheat. A, upper part of stem with leaves and flowers, × ⅕. B, leaf, × ½. C, flowers in various stages, × 3. D, fruit, × 4. E, diagrammatic transverse section of fruit. *e,* endosperm. *eb,* embryo. *o,* ochrea. *p,* pericarp. *t,* testa.

Buckwheat is usually grown for the achenes, which are milled into flour for human consumption, or used for stock-feeding. It is generally a less satisfactory flour-crop than the true cereals, and is normally only used when these cannot be grown. Its advantages are its quick growth and its tolerance of low fertility and of soil acidity. Its disadvantages (in addition to the very high husk content) are its very weak stem, which is likely to lodge on even moderately fertile soils, and its uneven ripening. In the true cereals flowering is confined to a week or two at most, and all fruits ripen at about the same time, but in buckwheat flowering is continuous over a long period. In consequence of this, some flowers are still in bud at the time when the earliest-formed fruits are mature. Moreover, the plant is still actively growing at the time when it is necessary to harvest it in order to avoid undue loss of this early seed, and drying of the cut crop is therefore difficult.

Buckwheat is virtually obsolete in Britain, and the small quantity of buckwheat flour used for culinary purposes is imported. It may very occasionally be used as a green-manuring crop for ploughing in, or sown in small patches to provide food for game. Some danger of stock-poisoning attaches to its use as green fodder. When grown for seed it is drilled at about 50 kg/ha, or at up to 150 kg for green manuring. Drilling must be in late spring, as buckwheat is not frost-hardy and does not germinate well at low temperatures. Yields of 1·5–2·0 t/ha may be expected.

Buckwheat probably originated in Central Asia, and was introduced into Europe during the Middle Ages. A number of distinct forms exist, including *Japanese,* with rather larger 'seeds', and *Silver Grey,* with pale grey pericarp, but *Common Grey,* the type already described, is the only one used in Britain. *Tartarian Buckwheat,* hardier, and with smaller, rounded achenes, is a distinct species, *Fagopyrum tataricum* (L.) Gaertn., cultivated in Asia. The name buckwheat means 'beech-wheat', from its use for flour and the resemblance of the achene to a miniature beech-nut.

### *Rheum rhabarbarum* L. (*R. rhaponticum* auct. non L.).     **Rhubarb**

A perennial with short rhizomes forming a thick, irregular root-stock, from which arise very large leaves, of which the petiole is eaten. Inflorescence a massive panicle 2 m tall, with very numerous small white hermaphrodite flowers. Perianth segments six, stamens from six to nine, ovary with three styles, maturing to form an oblong, slightly winged achene. Usually propagated vegetatively by splitting of rootstock, and named varieties are thus clones, but can be grown from 'seed' (nuts or achenes); this is the normal way in which new cultivars are produced. Grown on a large scale either in open ground, or lifted roots 'forced' in the dark after dormancy has been broken by frost or after treatment with gibberellic acid, to give earlier, more tender, brighter-red etiolated petioles. Of Central Asian origin; other species such as *R. palmatum* L. (with deeply-lobed leaves) possibly involved in development of modern cultivated varieties. Not commonly grown until early nineteenth century. Leaf blades not eaten, often poisonous owing to high oxalic acid content.

The genus *Rumex,* with inner perianth segments much larger than the outer, includes the important weeds, docks: two species are occasionally cultivated in private gardens. These are *Rumex acetosa* L. (of which the wild form is the weed common sorrel), garden sorrel, a sharp-tasting broad-leaved perennial salad plant, and *R. patientia* L., herb patience or sorrel dock, a perennial spinach-like leaf vegetable for spring use.

### CANNABACEAE

*General importance.* Only one species, *Humulus lupulus,* the hop, is of importance in British agriculture, and that as a localized crop in the hands of specialist growers. Hemp is of some importance as a fibre and drug plant in warmer countries, but its cultivation is now obsolete in Britain.

*Botanical characters.* A very small family of two genera only, with inconspicuous unisexual flowers, formerly united with either *Urticaceae* or *Moraceae,* but sufficiently distinct to merit treatment as a separate family. Plants herbaceous, with stipulate, palmately-lobed or divided leaves. Dioecious, the male plants with axillary panicles, the female with sessile flowers in dense, spike-like clusters. Male flowers with five perianth segments and five stamens. Female flowers with small, entire cup-like perianth, and a single, one-seeded ovary with two large stigmas. Bracts and bracteoles conspicuous, surrounding female flowers. Fruit an achene, seed endospermic, embryo coiled.

### *Humulus lupulus* L.   **Hop**

A perennial climbing plant grown for the female inflorescences which are used in brewing. Annual stems arise in spring from an underground root-stock, and may climb by twining (clockwise) around a string or other support to a height of some 6 m. The stems are angular,

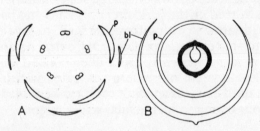

Fig. 82. *Cannabaceae.* Floral diagrams of A, male flower; B, female flower. *bl,* bracteole, *p,* perianth.

hollow, and hairy with stiff reflexed hairs; they bear opposite palmately-lobed leaves of rather variable shape, with pointed scarious stipules at the base of the petiole. Male plants produce in summer branched axillary and terminal panicles of loosely-arranged flowers 4–5 mm in diameter, consisting of five blunt, lanceolate perianth segments and five anthers on short filaments. The female plants produce smaller axillary panicles of hop-cones, or *strobili*. Each strobilus consists of a short axis (the 'strig') on which are borne a number of spirally-arranged bracts, or strictly of paired structures representing the stipules of the bracts. The bract lamina is only very occasionally developed in abnormal inflorescences. In the axil of each pair of stipular bracts is a very short, branched axis bearing four flowers each subtended by a bracteole. Each flower consists of a single ovary surrounded by the cup-like perianth. Two large papillose stigmas are present which reach a length of about 4 mm when receptive and stand out well beyond the bracts and bracteoles, giving a characteristic brush-like appearance to the strobilus at this ('burr') stage.

The stigmas receive wind-borne pollen from male plants, where these are present; rapid growth of the bracts and bracteoles takes place even in the absence of pollination, so that the appearance of the strobilus changes and becomes fir-cone-like. When mature, the strobilus is some 4 cm long, the individual stipular bracts and

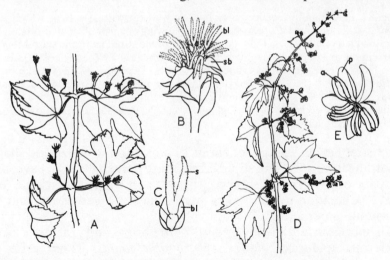

Fig. 83. Hops. A, part of female plant with leaves and young strobili, × ½. B, strobilus at 'burr' stage, × 2. C, single female flower at same stage, × 4. D, upper part of a stem of male plant in flower, × ½. E, single male flower, × 4. *a*, anther. *bl*, bracteole. *p*, perianth segment. *s*, stigma. *sb*, stipular bract. *o*, ovary.

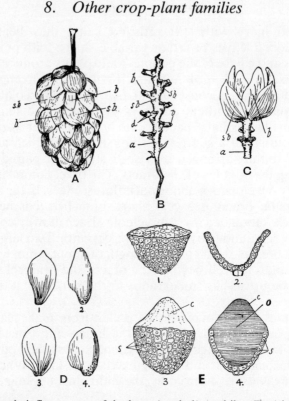

Fig. 84. Female inflorescence of the hop. A, whole strobilus. B, strig, with bracts and bracteoles removed. C, part of strobilus, showing one pair of stipular bracts and the associated bracteoles. *a,* strig, the main axis of the strobilus. *d,* the short branch-system in the axil of the undeveloped bract. *sb,* stipular bract. *b,* bracteole. (In B, *sb* and *b* indicate the scars left by the removal of these structures.) D, 1, stipular bract and 2, bracteole of the variety Fuggles; 3 and 4, of the variety Bramling. E, lupulin glands, × about 150; 1, side view and 2, vertical section of young gland; 3 and 4, of mature gland. *s,* secretory cells of gland. *c,* cuticle. *o,* resinous secretion. (From Percival, *Agricultural Botany.*)

bracteoles having a length of about 15 mm, varying in size and shape according to variety. The stipular bracts are only slightly concave, but the base of the bracteole curves round to envelope the achene, or, where pollination has not taken place, the shrivelled ovary, and its surrounding perianth.

The perianth and the lower parts of the bracts and bracteoles bear numerous epidermal glands, the *lupulin glands.* These arise as shortly-stalked, cup-like structures one cell in thickness. At about the time of pollination the cells of the cup start secreting a yellow resinous fluid, which accumulates under the cuticle covering the inside of the cup. The secretion of resin continues during the growth of the

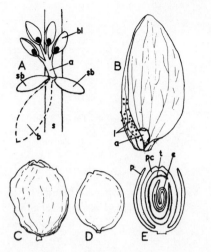

Fig. 85. Hops. A, diagram to show relation of stipular bracts to flowers in female strobilus. *a*, short axis in axil of bract, *b*, of which only the stipules, *sb* (stipular bracts), are developed. *bl*, bracteole subtending flowers. *s*, 'strig', or main axis of strobilus. B, bracteole with mature fruit, × 2½. C, achenes surrounded by cup-like perianth, × 7. D, with perianth removed. E, diagrammatic vertical section of C. *a*, achene. *l*, lupulin glands. *p*, perianth. *pc*, pericarp. *t*, testa. *e*, embryo.

strobilus, until the cuticle becomes strongly convex, and the gland with its enclosed resin almost spherical. The resin solidifies and becomes somewhat opaque so that the glands appear like a dusting of fine sulphur over the surfaces of the mature strobilus.

It is for these lupulin glands that the hop plant is grown. Their secretion, lupulin, is a complex mixture of resin and resin-like substances which not only impart to beer its bitter taste, but also act as an efficient preservative, by preventing rapid bacterial growth. Lupulin contains $\alpha$ and $\beta$ bitter acids, of which the former are the active ingredients, and the $\alpha$-acid content is a good measure of hop quality. Lupulin extracts may be used in brewing instead of the whole strobili.

Fruit an achene, dark brown, 2·5 mm, partially covered by persistent cupule. Sown only for the production of new varieties. Germination epigeal, cotyledons blunt-lanceolate, first foliage leaves ovate, coarsely-toothed, not lobed. A short, thickened tuberous root-stock is produced during the first year of growth and from the crown of this the aerial stems arise.

*Cultivation.* Since the strobili are the parts of the plant which are of value, only female plants are grown, except that, where seeded hops are required, a small proportion (about 1 in 200) of male plants is included. This has been the general practice in Britain, but accession to the E.E.C. has necessitated a change to the production of seedless hops. This involves the elimination of male plants (including wild male hops in hedges, etc.) and results in a reduction of some 30% in total hop yield. Since, however, the achene is in itself of no value, its

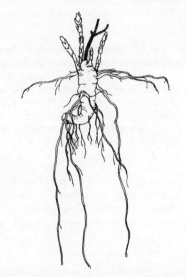

Fig. 86. Young hop plant grown from seed; beginning of second year, $\times \frac{1}{2}$.

loss is not necessarily a disadvantage, and in suitable varieties the useful $\alpha$-acid yield is little reduced by the absence of pollination.

Hops are propagated vegetatively; the aerial twining stems die at the end of the year, but the base of the stems below ground remains alive, and these stem bases are used as cuttings. Each cutting (sett) consists of a 10 cm length of stem-base removed from the parent plant; it bears buds and readily produces adventitious roots. Cuttings are placed in a nursery bed and planted out the following year in their permanent positions. The usual spacing is about 2 m square, depending on variety, and plants remain productive for many years. Hops require good soil, heavy manuring, and shelter from winds and a large expenditure is involved in the provision of a wire framework to support the strings up which they grow, and in the stringing, training and removal of excess stems and spreading rhizomes, as well as cultivations to destroy weeds and spraying to control insect pests and fungus diseases. Harvesting, formerly by hand-picking of the mature strobili and therefore requiring a very large amount of seasonal labour, has become increasingly mechanized. Mechanical picking differs from hand-picking in that the stems ('bines') must be cut and fed to the machine, so preventing the translocation of remaining food material from the leaves and stems to the rhizomes.

The picked hops must be artificially dried in kilns (oast-houses) and exposed to sulphur dioxide as a bleaching and preservative agent. During drying and subsequent packing careful handling is necessary to avoid loss of the dry lupulin glands.

### Origin and range of types

Hops are probably native in Britain, although many of the hop plants found wild are escapes from cultivation, but the use of this species in brewing was introduced in the sixteenth century from continental Europe. It is probable, therefore, that the cultivated forms are derived from continental rather than British plants. Since the hop is normally propagated vegetatively, the varieties are strictly speaking clones. New varieties are produced from seed —that is, by sowing the achenes; about half the resultant plants are males which must, of course, be discarded. Breeding work is necessarily unusually difficult in a dioecious plant such as hops. Certain characters, such as vigour of growth and disease resistance, are visible in the male parent, but there is no method except some form of progeny testing, of assessing its genetical make-up as far as yield and quality are concerned. Nevetheless, considerable progress has been made, notably at Wye College, Kent, in the development of new and improved varieties.

*Cultivars.* In the 1950s some 75% of the British hop area was occupied by the variety Fuggles, a mid-season wilt-susceptible hop with narrow bracteoles; it is of rather low quality, but adapted to a wide range of soil conditions. Golding, earlier and of higher quality, with broad bracteoles, was grown on the better soils. The rate of change of varieties in a perennial crop is of course comparatively slow, but Fuggles declined steeply in popularity during the 1960s and 70s, being replaced by newer cultivars, mainly developed at Wye, with better disease resistance and higher $\alpha$-acid yields. These included Northern Brewer, which performs well when grown as a seedless hop. During the 1970s provision was made for the setting up of seedless hop areas, free from male plants. Triploid cultivars are utilized in New Zealand, but these are not completely sterile, and produce seeds in excess of the 2% permitted under E.E.C. regulations.

### *Cannabis sativa* L.     **Hemp**

Hemp is a tall-growing, slender, rather tender annual, with large alternate leaves palmately divided into lanceolate, coarsely-toothed, pointed leaflets. Like hops, it is dioecious, the male plants bearing small axillary panicles of yellowish-green flowers, the female axillary leafy clusters not forming a compact strobilus. The male flowers are similar to those of hops; the female flowers have more slender

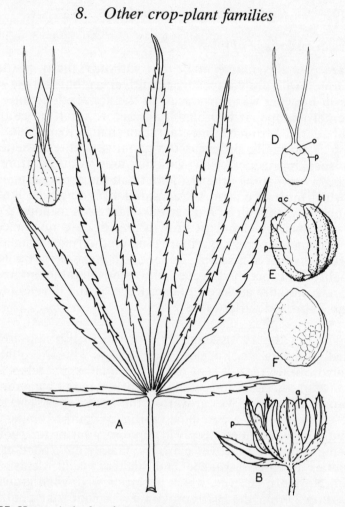

Fig. 87. Hemp. A, leaf, × ½. B, male flower, × 4. C, female flower enclosed in bracteole; D, with bracteole removed. E, ripe achene surrounded by remains of perianth and bracteole. F, achene alone, × 4. *a,* anther. *ac,* achene. *bl,* bracteole. *o,* ovary. *p,* perianth.

stigmas and are closely surrounded by the inrolled tube-like brac-teole. The fruit is a greyish-brown achene, about 4 mm long (about 75 000/kg) used to some extent as bird-seed.

Hemp was formerly widely cultivated in Britain for its fibre; this, like that of flax (p. 196), is derived from the groups of fibre cells which occur on the outer side of the vascular bundles of the stem. The fibre cells differ from those of flax in being stouter and more lignified, and the extracted fibre, obtained by retting and scutching in the same way as with flax, is used for coarse fabrics and ropes. The cultivation of hemp somewhat resembles that of flax, 'seed' (achenes) being drilled or broadcast at about 100 kg/ha in April or May. The resultant crop consists of about equal numbers of male and female plants, and since the male plants mature earlier than the female, the best quality fibre is only obtained by hand-pulling the male plants first and then harvesting the female plants separately a month or more later. Yields of about 600 kg/ha of fibre might be expected, together with a similar weight of seed. Hemp is thus an expensive crop to grow, and is not now economic in Britain owing to the competition of imported fibres, including not only true hemp but also the technically similar mono-cotyledonous leaf-fibres, Manila hemp (from the banana-like *Musa textilis*), sisal (from *Agave sisalana*) and New Zealand hemp (from *Phormium tenax,* which has also been experimentally grown in south-western England).

In hot climates the glandular hairs of the hemp plant secrete resins containing narcotic alkaloids, and it is thus an important and danger-ous drug plant (hashish, marijuana), controlled in Britain by the Misuse of Drugs Act 1971.

### JUGLANDACEAE AND RELATED FAMILIES

A number of families of woody plants with small inconspicuous flowers, often arranged in catkin-like inflorescences, may be men-tioned as including some crops which can be grown in Britain for their edible fruits or seeds.

In the *Juglandaceae* the male flowers are in catkins, the female flowers solitary or in small groups. In *Juglans regia* L., the walnut, the two joined carpels develop to give a drupe-like fruit. The mesocarp is not edible, the endocarp forms the woody shell, and the seed is the part eaten, with high oil and protein content. Walnuts originated in south-eastern Europe, and are grown only to a very limited extent in Britain. In the *Corylaceae* the male flowers are again in catkins; in *Corylus avellana* L., the hazel nut, the female flowers each develop to form a true nut, surrounded by a thin involucre formed from bracts. The woody shell is formed from the whole pericarp, and the edible part is the seed, with high protein and oil content. Grown on a commercial scale in south-eastern England; cultivars are clones (cobs and filberts) propagated by rooted suckers. In the *Fagaceae,* which includes the important timber trees oak and beech, *Castanea sativa* Mill., the sweet chestnut, has the female flowers in small groups, and the resultant group of nuts (a single nut in large-fruited edible forms) is enclosed in a spiny four-valved involucre formed from bracts. Pericarp leathery, seed with high starch and low oil and protein content, and therefore eaten cooked. Of southern European origin, chestnuts used in Britain mainly imported.

In the *Moraceae* the flowers are densely aggregated, and the whole inflorescence may become succulent, to form a multiple fruit. In *Morus nigra* L., the black mulberry, a monoecious tree of Chinese origin, the small bracts surrounding the female flowers become succulent, so that each female inflorescence develops into a multiple fruit resembling a blackberry in appearance. In *Ficus carica* L., the fig, of southern European origin, very numerous small flowers are borne on the inner surface of the very deeply concave, almost closed, inflorescence receptacle. It is this receptacle which becomes succulent, the true fruits being the pips.

### BORAGINACEAE

*General importance.* Contains only one agricultural crop, Russian comfrey, which has been recommended as a forage crop, but is not in general cultivation. Named from genus *Borago,* borage.

*Botanical characters.* Plants herbaceous, usually rough-hairy; leaves simple, alternate, exstipulate. Inflorescence a monochasial cyme. Flowers actinomorphic, insect-pollinated. Calyx of five joined sepals; corolla tubular or with rotate limb, five-lobed, mouth of corolla-tube partly closed by scales. Stamens five, epipetalous. Gynaecium superior, of two united carpels each divided into two lobes, giving four single-seeded nutlets around the central style.

### *Symphytum* × *uplandicum* Nyman. (*S.* × *peregrinum* Ledeb.) Russian Comfrey

A hybrid, or more strictly a hybrid swarm (i.e. a series of hybrids together with their back-crosses with both parents), between *S. officinale* L., common comfrey, and *S. asperum* Lepech. (*S. asperrimum* Donn.), prickly comfrey.

A perennial, with stout, somewhat tuberous roots, forming a series of branching crowns. Radical leaves hispid, broadly-lanceolate, stalked. Stems erect, 1·5 m high, branched, bearing similar but smaller sessile leaves, variably decurrent. Flowers pendulous in branched cymes at top of stems; calyx with pointed teeth, corolla broadly tubular, usually dull pinkish-purple. Mature nutlets rarely produced, 4–5 mm, rugose, black.

*S. asperum* has smaller leaves, not decurrent, and bright blue flowers with blunt calyx-teeth; *S. officinale* is usually a lower-growing

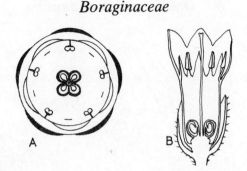

Fig. 88. Floral characters of *Boraginaceae*. A, floral diagram. B, vertical section of flower.

plant, with leaves strongly decurrent, dull purple or white flowers, and smooth black nutlets. The form of Russian comfrey now grown is nearer to *S. officinale* in appearance than to *S. asperum*; it is reported to have been introduced from Russia about 1870, but crosses between the wild British *S. officinale* and the earlier-introduced *S. asperum* (formerly used as crop-plant, now sometimes grown as garden ornamental, and occasionally naturalized) may have contributed to the present range of forms, which includes several different clones.

A quick-growing plant producing a large bulk of foliage; not palatable in the fresh state and not tolerant of hard grazing. The treatment recommended is cutting and feeding wilted or as silage; with five to eight cuts per year, annual yields of 125–250t/ha fresh weight have been reported from established stands under good conditions. With a dry-matter content of about 12%, of which 20–25% is crude protein, this yield compares favourably with that of other, more commonly grown crops. That comfrey has not become established as a common crop may be attributed partly, perhaps, to the very considerable confusion which has surrounded the plant, and the fact that many of the trials carried out have been with *S. asperum* or other lower-yielding forms, but still more to the inherent disadvantages of the plant as a general farm crop. It can only be propagated vegetatively; this involves the planting-out of root-cuttings or divided crowns in autumn or spring (the plant dies down to ground level in winter) at about 12 500/ha and a consequent very high cost of establishment. Row-crop cultivation is necessary to control weeds, and a high level of fertility must be maintained; maximum yields are not attained in the first year or two, and the plant must be treated as a long-duration crop. Costs of utilization involving frequent cutting are likely to be high; further, the plant, once established, is often difficult to eradicate completely when this is required, and may persist as a weed.

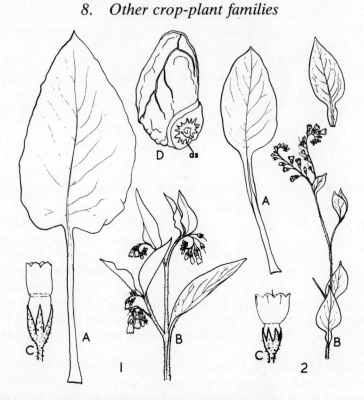

Fig. 89. 1, Russian comfrey (*Symphytum* × *uplandicum*). 2, Prickly comfrey (*S. asperum*). A, basal leaf, × ¼. B, upper part of flowering shoot, × ¼. C, flower, × 1. D (one only), nutlet (rarely formed), × 7. *as,* attachment surface.

The reported presence of carcinogens in comfrey is likely to reduce its use still further.

## LABIATAE

*General importance.* Includes a number of strongly scented plants, mainly of Mediterranean origin, of which a few are grown as culinary herbs for flavouring.

*Botanical characters.* Stems square, leaves decussate. Flowers in axillary clusters often appearing whorled. Calyx of five united sepals, corolla tubular, four- or five-lobed, zygomorphic, often two-lipped (whence the family name). Stamens usually four, epipetalous. Ovary superior of two joined carpels, splitting into four single-seeded nutlets resembling those of the *Boraginaceae*.

*Mentha,* the mint genus, includes numerous species and interspecific hybrids; corolla almost equally four-lobed, perennial by rhizomes, vegetatively propagated. *Mentha spicata* L., spearmint, probably of hybrid origin, and with sessile leaves and slender inflorescence, is the form commonly grown for flavouring and for mint sauce. *M.* × *piperita* L., peppermint, a sterile hybrid between *M. aquatica* L. and *M. spicata,* has been grown on a field scale for the production of peppermint oil by steam distillation.

*Salvia officinalis* L., sage, has large strongly two-lipped flowers with two stamens only, leaves *c.* 5 cm long, grey-green, rugose; plant perennial, woody at base, without rhizomes. Leaves used for flavouring.

*Thymus vulgaris* L., thyme, a small woody perennial with sessile leaves *c.* 6 mm long, with characteristic scent, and small pale purple flowers, is also commonly grown as a culinary herb.

Other members of the family which can be used for flavouring, but which are rarely grown on a commercial scale in Britain include *Origanum majorana* L., sweet marjoram, a tender perennial grown as an annual; *O. onites* L., pot marjoram, hardier and with sessile leaves; *Ocimum basilicum* L., sweet basil, annual, and the similar but smaller *O. minimum* L., bush basil; *Satureja hortensis* L., summer savory, annual, and *S. montana* L., winter savory, perennial, and also *Melissa officinalis* L., lemon balm, perennial. *Lavandula angustifolia* Mill. (*L. vera* DC), lavender, and *Rosmarinus officinalis* L., rosemary, are evergreen shrubs used in perfumery; essential oil extracted by steam distillation.

*Stachys affinis* Bunge, Chinese artichoke, is an uncommon 'root'-vegetable producing numerous tubers on short rhizomes in the same sort of way as the Jerusalem artichoke. Tubers consist of some five or six swollen internodes, constricted at the nodes; about 20% dry matter, yielding at the rate of perhaps 12 t/ha; difficult to lift and store and grown only on a small private garden scale.

## DIPSACACEAE

*General importance.* Includes only one crop-plant, teasel, of which a small area is grown, the dry inflorescences being used in the finishing of woollen cloth.

*Botanical characters.* A small family of herbaceous plants, with opposite leaves and rather small flowers usually massed together in a compact head, often somewhat resembling that of the *Compositae* (see below, p. 227), but readily distinguished by the free, not joined, anthers.

### *Dipsacus sativus* (L.) Honkeny.   **Cultivated Teasel**

A prickly biennial with long tap-root, producing in the first year a rosette of large broadly-lanceolate leaves, and in the second year an erect, much-branced stem to 1·5 m high. Stem leaves connate, so that the bases of the two opposite leaves at each node join to form a cup. Branches terminated by dense, erect, capitulum-like flower heads, about 8 cm long by 4 cm in diameter. At the base of each head is a whorl of long, spiny bracts forming the *involucre,* and above this the swollen axis (*receptacle*) bears numerous shorter *receptacular bracts.* In the axil of each of these latter bracts is a single flower surrounded at its base by a cup-like *involucel* formed from fused bracteoles. Each flower consists of an inferior ovary enclosed in the involucel and surmounted by a short, four-toothed calyx and a tubular, four-lobed mauve corolla, on which are borne the four free stamens. The inferior ovary, which is composed of two united carpels, contains a single pendulous anatropous ovule, and bears a single style with undivided stigmatic tip. Cross-pollinated by bees and flies. Fruit an achene, about 5 mm long, with persistent involucel and calyx.

In the cultivated form, fuller's teasel, the bracts of the involucre are comparatively short and spread horizontally, and the receptacular bracts are stiff and recurved. It is these receptacular bracts which are the effective part of the inflorescence when it is dried and mounted on roller frames and used to raise the knap of woollen cloth. In the wild teasel, *Dipsacus fullonom* L., the involucral bracts are longer and curved upward around the head, and the receptacular bracts are softer and straight, and therefore useless for cloth-finishing. The two forms are sometimes treated as subspecies, in which case

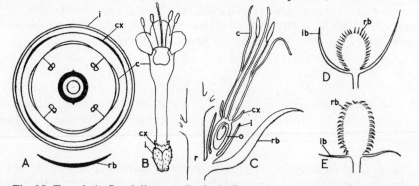

Fig. 90. Teasel. A, floral diagram. B, single flower in ventral view. C, in diagrammatic vertical section. D, diagram to show general shape of flower-head of wild teasel; E, of cultivated teasel. *i,* involucel. *cx,* calyx. *c,* corolla. *o,* inferior ovary, *r,* receptacle (main axis of flower-head). *rb,* receptacular bract. *ib,* involucral bract.

the cultivated teasel is *D. fullonum* L. subsp. *sativus* (L.) Thell., and the wild form subsp. *fullonum*.

Teasels are cultivated in small areas in various parts of the temperate regions, in England on a small scale in Somerset. They are a crop which requires a large amount of hand labour. 'Seed' (achenes) sown in April, often in special seed-beds, and young plants transplanted after shortening of the tap-root and spaced at about 32 000/ha. They bloom in July of the following year, and individual heads are cut by hand as they become ready. Yields of some 400 000 heads/ha may be obtained; very careful drying and packing of the heads is necessary to avoid damage to the bracts. 'Seed' is obtained from a proportion of heads left to become fully ripe.

### COMPOSITAE

*General importance.* The *Compositae* provides no crops of outstanding importance in British agriculture; chicory and yarrow are herbs sometimes included in leys, and sunflowers are sometimes grown as an arable crop. Among horticultural crops, lettuce is of greatest importance.

*Botanical characters.* The largest of all families of flowering plants, including about one-tenth of all known species. Members of the family are world-wide in their distribution, and occur in almost every possible type of habitat.

The habit is very varied, including trees, shrubs and herbaceous plants of many types. Leaves are exstipulate, usually alternate, very variable in shape, but often pinnately cut or divided. The most characteristic feature of the family is the inflorescence, which is always a *capitulum*. Each capitulum consists of an enlarged stem-structure, the *receptacle,* which is surrounded by one or more whorls of bracts and bears on its upper surface a number of sessile individual flowers. The bracts around the receptacle form the *involucre*; the individual flowers, which may be subtended by scales arising from the surface of the receptacle, are usually known as *florets** on account of their small size compared with that of the whole capitulum. The receptacle may be flat, convex or cylindrical, and may bear from a few (in certain genera, one only) to many hundred florets.

Each individual floret has an inferior ovary from which the other

---

* It should be noted that *floret* is used in a different sense in describing the *Gramineae,* where it refers to an individual flower together with its associated lemma and palea.

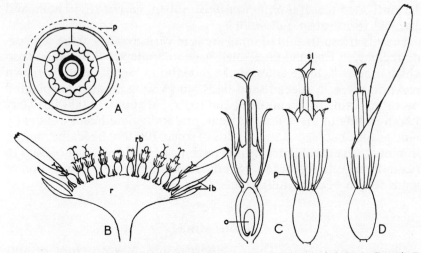

Fig. 91. Floral characters of *Compositae*. A, floral diagram (of single floret). B, diagrammatic vertical section of capitulum. C, vertical section and side view of tubular floret. D, side view of ligulate floret. *a,* anther tube. *o,* ovule. *p,* pappus. *r,* receptacle. *rb,* receptacular bract. *ib,* involucral bracts. *sl,* stigmatic lobes.

floral parts arise. The calyx is either absent or replaced by a ring of hairs (or, in some genera, scales) known as the pappus. The corolla is formed by five united petals, and on it are borne five stamens which are united by their anthers to form a tube surrounding the single style. The inferior ovary is formed from two united carpels, and contains a single basal anatropous ovule. The appearance of the floret largely depends on the shape of the corolla, and the two distinct types are recognizable: (1) *tubular florets,* with an actinomorphic cylindrical or bell-shaped corolla, divided at the top into five broad lobes; and (2) *ligulate florets,* in which only the base of the corolla is tubular, and the remainder forms a flat, strap-like or lanceolate structure, often with five minute teeth at the apex. The capitulum may consist entirely of tubular florets (as in thistles) or entirely of ligulate florets (as in dandelions), or may have a central *disk* of the tubular florets surrounded by a single ring of ligulate florets forming the *ray* (as in daisies).

Cross-pollination is by insects, and the whole capitulum forms a conspicuous attractive structure, in spite of the small size of the individual florets. If insect-pollination does not take place, it is still possible for selfing to occur, and in some forms self-pollination is the general rule. The style is terminated by two stigmatic lobes, of which only the inner surface is receptive. Dehiscence of the anthers is introrse, and pollen is shed into the tube formed by the united anthers. The two stigmatic lobes are not separated when the style grows up through the anther tubes, and pollen is therefore deposited only on their outer non-receptive surface. Later they separate so that the inner surfaces of the stigmas are in a position to receive pollen from visiting insects. Still later they curve downwards, so that the receptive surfaces make contact with the pollen earlier deposited on the outer surface of the style, and self-pollination takes place if crossing has failed.

The inferior ovary develops into a single-seeded, dry, indehiscent fruit usually called an achene (since it is derived from two carpels, and is inferior, the special term *cypsela* is sometimes used). The pappus, if present, usually remains attached to the achene, and forms a very efficient means of wind dispersal. The seed is almost non-endospermic, with a large, straight embryo, usually containing oil as a food-reserve.

The large size of the family, and the very uniform structure of the inflorescence and flower, make classification difficult, and the *Compositae* is usually divided into numerous tribes, separated by very small differences. It is sufficient here to note the two main divisions or sub-families, *Tubiflorae* and *Liguliflorae*.

## TUBIFLORAE

Some or all of the florets tubular; oil canals often present, but no latex.

### *Helianthus annuus* L.   Sunflower

A tall annual, with large, rough-hairy, cordate leaves, alternately arranged on the stout, little-branched stem, and very large terminal capitula. The receptacle is flat or slightly convex, with several rows of large, pointed, involucral bracts. Florets subtended by broad scale, corolla yellow; disk florets very numerous, tubular, fertile; ray florets ligulate, large, sterile. Achenes broad, angular, large (c. 10 mm); pappus scaly, not persistent. Seedling with long-oval, short-stalked cotyledons; first two leaves opposite, lanceolate, almost sessile, hairy.

Originated in North America, where wild and weed races are widespread. Cultivated forms are little branched, with large capitula, varying from giant Russian types 2 m and more tall with capitula 50 cm across to dwarf forms growing to 1 m; typical oil seed cultivars are about 1·5 m, with 25 cm capitula; achenes black, white or striped. A considerable amount of cross-pollination may take place.

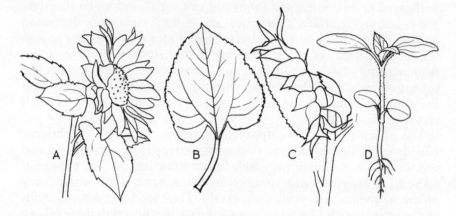

Fig. 92. Sunflower (Mars). A, small lateral capitulum, in flowering stage, ×¼. B, leaf, × ⅛. C, capitulum at ripe achene stage, × ⅛. D, seedling, × ¼.

Sunflowers are an important oil-producing crop in southern and south-eastern Europe. Modern varieties have up to 48% oil in the achenes, and with yields of from 2 to 4 t/ha are potentially a useful crop in England. Germination is slow below 7°C and sowing can not normally take place before early April, early varieties, which alone are suitable here, coming to harvest in September. Drilled to give 50 000–60 000 plants/ha, harvested with high-set combine with modified cutter-bar, dried at not over 50°C to 9% moisture for storage. Pre-harvest bird damage may be severe on small areas.

Sunflower oil, used as an edible oil and for margarine manufacture, has a high content of unsaturated fatty acids, with up to 68% linoleic acid in samples from cool areas; under warm conditions the oleic acid content is higher, and the linoleic lower. The residue after oil extraction is a useful feed; decorticated sunflower cake may contain up to 50% protein with a high methionine content.

The older open pollinated cultivars are being replaced by F.1 hybrids derived from male-sterile female parents. Some French cultivars are available which are earlier maturing, higher yielding and with higher oil content than the Canadian varieties such as Mars and Pole Star which were used to some extent in England during the 1940s, and are therefore more promising for use here.

Sunflowers could form a useful break crop in cereal growing areas; their use as an oil seed crop might be expected to be most successful in the areas where grain maize gives satisfactory yields, but sunflower silage, for which dry matter yields of 8·5 t/ha have been recorded, would be a possibility over a much wider area.

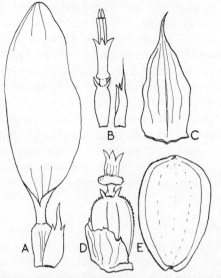

Fig. 93. Sunflower (Mars). A, ligulate floret and receptacular bract, × 1½. B, tubular floret and bract, × 1½. C, involucral bract, × ½. D, tubular floret with nearly mature achene, × 1½. E, mature achene, × 3.

## *Helianthus tuberosus* L.    Jerusalem Artichoke

Similar to sunflower, but perennating by tubers; stems more branched, leaves mainly opposite, with broad, winged petiole; capitula small, rarely produced in Britain; vegetatively propagated by the tubers. Named artichoke from the supposed resemblance in flavour of the tubers to the globe artichoke (*Cynara*, see below). Jerusalem is apparently a corruption of either girasole, the Italian name for sunflower, or of Ter Neusen, in Holland, from whence the crop may have entered Britain; it is of North American origin, and has no connection with Palestine. The French name topinambour is almost equally inappropriate, being derived from that of a Brazilian tribe with whom the plant was mistakenly associated at the time of its introduction into Europe at the beginning of the seventeenth century.

Grown primarily for its tubers, which are more irregular in shape than those of the potato, borne on shorter rhizomes, and with larger persistent scale-leaves subtending the buds; size up to about 10 × 5 cm, with a mass of perhaps 100 g. Internally the tubers consist mainly of parenchyma, partly primary tissue produced by proliferation of the pith cells, partly secondary tissue produced by the cambium. The well-marked cork layer found in potatoes is absent, and tubers wilt rapidly during storage. Since they are frost-hardy, they are usually left in the ground over winter and lifted as required. The food reserve in the tubers is mainly inulin, not starch; they contain 16–20% dry matter, of which some 7% is protein. Artichokes are mainly grown as a vegetable for human consumption but, being fairly tolerant of low fertility and poor cultivations, they are sometimes recommended as a farm crop on difficult land. Rarely grown except on a small scale; tubers planted in early spring at a rate equivalent to about 3 t/ha to give some 30 000 plants/ha; yield may be in the region of 25–30 t/ha; difficult to lift cleanly and volunteer plants may persist.

Although artichokes are vegetatively propagated, few distinct forms exist, and cultivar names are not commonly used; the type mainly grown in Britain has white-skinned tubers, but purple-skinned forms exist.

*Achillea millefolium* L., yarrow, a low-growing rhizomatous perennial with finely divided dark green leaves and small white capitula in corymbs, is occasionally used as a constituent of leys, or of special herb-strip mixtures for grazing, on account of its high mineral content. The 'seed', consisting of the very small, flat, silvery-white achenes (1·8 mm, *c*. 11 million/kg) is expensive, and if used is sown at a rate of not more than 500 g/ha. *Chrysanthemum cinerariifolium* (Trev.) Vis., pyrethrum, is grown in warm temperate areas for its flowering capitula, from which the insecticide pyrethrum is extracted. *Cynara cardunculus* L., globe artichoke, is a tall thistle-like perennial, with large blue-flowered capitula, with all florets tubular. It is grown on a commercial scale in France, and to some extent in private gardens in Britain, as a vegetable, the fleshy receptacle and involucral bracts being eaten. Cultivars are clones, propagated by offsets. Cardoon is a rarely-grown form of the same species in which the young etiolated shoot is used. *Artemisia dracunculus* L., tarragon, is a herbaceous perennial up to 60 cm tall, with lanceolate leaves 3 cm long, vegetatively propagated and occasionally grown as a flavouring herb. Safflower and niger are annual warm-temperate oil-seed crops producing edible drying oils of high linoleic acid content. *Carthamus tinctorius* L., safflower, is spiny, with bright orange florets, all tubular, and white achenes without pappus, with 24–36% oil. *Guizotia abyssinica* Cass., niger, has yellow disk and ray florets and large black achenes without pappus, with 38–50% oil.

LIGULIFLORAE

All florets ligulate; latex vessels present throughout the plant, associated with phloem.

## *Cichorium intybus* L.   **Chicory**

A perennial with stout tap-root and a rosette of large, somewhat hairy leaves with pointed, shallow lobes, Erect annual stems are produced, growing to about 1·5 m high, and bearing similar but smaller leaves, and axillary clusters of almost sessile, few-flowered, blue capitula, 3–4 cm in diameter. Achenes angular, pale brown, rather irregular in shape, about 3 mm long, *c.* 650 000/kg with a pappus of short, finely-divided scales.

A common wild plant of chalk and limestone soils in England. Cultivated forms used in three distinct ways:

(1) As a grassland herb in the same way as yarrow. High-yielding and usually palatable, but rather difficult to manage; not persistent under hard grazing, and may become coarse and unpalatable under lenient grazing or hay. Sown at rates up to 2 kg/ha.

(2) As a root-crop, in the same sort of way as sugar beet, the roots being processed for the manufacture of a coffee substitute. The form

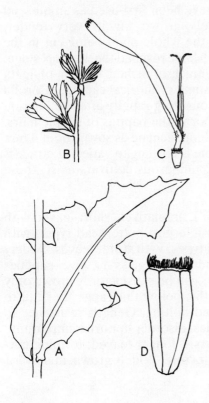

Fig. 94. Chicory. A, leaf, × ½. B, capitula, × ½. C, single floret, × 2. D, mature achene, × 10.

used is the large-rooted Magdeburg chicory (which is also the form. usually supplied for use (1) above), drilled at about 5 kg/ha.

Yields of 25 t/ha of washed roots are obtained, but lifting of the long roots is difficult, and roots incompletely removed may develop adventitious shoots, and the plant persist as a weed.

(3) As a vegetable, the young etiolated leafy shoot being eaten. The form usually used is Witloof, resembling Magdeburg, but with broader leaves.

*Cichorium endivia* L., endive, is a very similar species with the leaves glabrous, less lobed, the stems shorter, and the achenes slightly larger, with longer scales. Grown as a leaf vegetable, often blanched, in either plain or curled-leaved forms. Usually annual or biennial.

### *Lactuca sativa* L.   **Lettuce**

Lettuce is not known wild, but is closely related to *Lactuca serriola* L., Mediterranean and also extending to Britain, and may well be derived from it. Lettuce is recorded as cultivated in Egypt in the fifth millenium B.C., and may perhaps have been first used as an oil-seed crop. It is an annual, now widely grown for its very tender, compactly-arranged entire leaves, the whole plant being cut in the vegetative condition. In plants which have reached the 'bolting' stage, the stem elongates, producing a much-branched system of leafy shoots up to 80 cm high, with very small cylindrical capitula. Corolla pale yellow; upper part of ovary elongating rapidly after pollination to form slender beak on which is borne the pappus of simple hairs. Beak and pappus removed in threshing; achene as sown about 4 mm long, flattened, ribbed, black or white according to variety. Lettuce is almost entirely self-pollinated and numerous cultivars exist; these can be grouped into:

(1) Cabbage lettuce (var. *capitata* L.), relatively short-leaved with compact flat or round heads. Divisible into butterhead types, with soft-textured leaves, and crisphead types, with stiffer leaves. Widely grown commercially outdoors and under glass to give supplies throughout the year; many cultivars can be satisfactorily grown only within a limited range of day-lengths, below this range they fail to heart, and above it they show rapid bolting. Genetic resistance to downy mildew and to the physiological disorder tipburn is important.

(2) Cos lettuce (var. *romana* Gars.), longer leaved, forming narrower upright heads, leaves crisp in texture. Widely grown, mainly for summer production.

(3) Leaf lettuce (var. *crispa* L.), forming a very loose head, from which the individual deeply lobed or curled leaves can be picked individually over a relatively long period. Not grown commercially.

(4) Celtuce (var. *angustana* Irish), a rapidly bolting form in which the stout succulent elongated stem is eaten. A minor private garden crop, of Chinese origin.

*Tragopogon porrifolius* L., salsify, is a biennial occasionally grown as a garden vegetable for its thickened tap-root. It has long, narrow entire leaves and, in the second year, large capitula on leafless, swollen peduncles. Corolla purple, achenes large (12 mm), with large pappus of compound hairs on long beak. *Scorzonera hispanica* L., black salsify, with shorter lanceolate leaves and somewhat smaller capitula of yellow flowers on leafy stems, achenes not beaked, is used in the same way.

# FURTHER READING

## PLANT STRUCTURE AND FUNCTION

Leopold, A. C. and Kriedman, P. E. *Plant Growth and Development,* 1975. McGraw Hill.

Cutter, E. G. *Plant Anatomy: I Cells and Tissues,* 2nd ed., 1978; *II Organs,* 1971. Arnold.

Esau, K. *Anatomy of Seed Plants,* 2nd ed., 1977. Wiley.

Hewitt, E. J. and Smith, T. A. *Plant Mineral Nutrition,* 1975. E.U.P.

Milthorpe, F. L. and Moorby, J. *An Introduction to Crop Physiology,* 1974. C.U.P.

Wareing, P. F. and Cooper, J. P. *Potential Crop Production,* 1971. Heinemann.


## CROP PLANTS: GENERAL

Hector, J. M. *Introduction to the Botany of Field Crops,* 1936. Central News Agency, Johannesburg.

Hayward, H. E. *The Structure of Economic Plants,* 1948. Macmillan.

Janick, J., Schery, R. W., Woods, F. W. and Ruttan, V. W. *Plant Agriculture,* 1970. Freeman.

Berrie, A. M. M. *An Introduction to the Botany of the Major Crop Plants,* 1977. Heyden.

Cobley, L. S. *An Introduction to the Botany of Tropical Crops,* 2nd ed., 1976. Longman.

Purseglove, J. W. *Tropical Crops: Dicotyledons,* 1974. Longman.

Herklots, G. A. C., *Vegetables in South East Asia,* 1972. Allen and Unwin.


## CROP PLANTS: ORIGIN AND HISTORY

Zukovsky, P. M. (transl. P. S. Hudson) *Cultivated Plants and their Wild Relations,* 1962. C.A.B.

Darlington, C. D., *Chromosome Botany and the Origin of Cultivated Plants,* 1963. Allen and Unwin.
Hutchinson, Sir Joseph (ed.) *Essays on Crop Plant Evolution,* 1965. Cambridge.
Hyams, E., *Plants in the Service of Man,* 1971. Dent.
Ucko, P. J., and Dimbleby, G. W. (eds) *The Domestication and Exploitation of Plants and Animals,* 1971. Duckworth.
Frankel, O. H. and Bennett, H. E. *Genetic Resources in Plants,* 1970. Blackwell.
Frankel, O. H. and Hawkes, J. G. *Crop Genetic Resources,* 1975. Cambridge.
Simmonds, N. W. (ed.) *Evolution of Crop Plants,* 1976. Longman.
Renfrew, J. *Paleoethnobotany,* 1977. Columbia.

### INDIVIDUAL CROP PLANTS AND FAMILIES

Vaughan, J. G., Macleod, A. J. and Jones, B. M. G. *The Biology and Chemistry of the Cruciferae,* 1976. Academic Press.
Nieuwhof, M. *Cole Crops,* 1969. Leonard Hill.
Kay, D. E. *Root Crops.* Crop Product Digests No. II, Tropical Products Institute.
Heywood, V. H. *The Biology and Chemistry of the Umbelliferae,* 1971. Academic Press.
Burton, W. G. *The Potato,* 1966. Veenman and Zonen.
Ivins, J. D. and Milthorpe, F. L. (eds) *The Growth of the Potato,* 1963. Butterworth.
Harris, P. M. *The Potato Crop,* 1978. Chapman and Hall.
Vaughan, J. G. *Structure and Utilisation of Oil Seeds,* 1970. Chapman and Hall.
Bolton, J. L. *Alfalfa,* 1962. Leonard Hill.
Lowe, J. *White Clover Research,* 1970. British Grassland Society.
Spedding, C. R. W. and Diekmahns, E. C. (eds) *Grasses and Legumes in British Agriculture,* 1972. C.A.B.
Hills, L. D. *Comfrey Past, Present and Future,* 1976. Faber.

### JOURNALS

Papers dealing with agricultural crop plants appear in a large number of journals, including:

*Journal of the National Institute of Agricultural Botany.*
*Journal of the British Grassland Society,* continued from 1979 as *Grass and Forage Science.*

*Euphytica.*
*Journal of Experimental Agriculture.*
*Journal of Applied Ecology.*
*World Crops.*
*Annals of Botany.*

Papers in these and other journals, including foreign publications, are abstracted in:

*Plant Breeding Abstracts.*
*Herbage Abstracts.*
*Field Crop Abstracts.*
*Horticultural Abstracts.*

Official notices relating to the National Lists of cultivars, and to additions to and deletions from these are published in the *Plant Varieties and Seeds Gazette*.

### LEAFLETS

Leaflets dealing with most of the major agricultural and horticultural crops are published by the Ministry of Agriculture, Fisheries and Food.

The National Institute of Agricultural Botany publishes classified lists of varieties of cereal and herbage crops which supplement the information in the *Plant Varieties and Seeds Gazette* by giving a brief indication of performance. The N.I.A.B. also issue a series of *Farmers Leaflets* and *Vegetable Growers Leaflets* covering most major crops. These leaflets, which are revised annually, give more detailed information on the better cultivars, and in many cases include Recommended Lists; they provide the best source of up-to-date information on cultivars. These and other publications are supplied free to Fellows of the N.I.A.B.; details of the Fellowship Scheme, and of the special scheme for students, may be obtained from The Secretary, N.I.A.B., Huntingdon Road, Cambridge.

# GLOSSARY

*Achene:* a small, dry, one-seeded indehiscent fruit.

*Actinomorphic:* symmetrical about more than one diameter; regular.

*Adnate:* united with another part of a different kind.

*Adventitious:* out of the ordinary course; applied, for example, to roots arising from a stem or buds arising from a root.

*Allopolyploid:* polyploid derived from more than one species.

*Amenity cultivar:* cultivar not intended for use as a productive crop; e.g. a lawn grass.

*Amphidiploid:* allopolyploid containing diploid complements of both parent species.

*Amylase:* enzyme catalysing breakdown of starch; includes α-amylase which attacks glucose linkages at any point in molecule.

*Amylopectin:* constituent of starch, with branched chains of glucose.

*Amylose:* constituent of starch, with unbranched chains of glucose molecules.

*Anatropous ovule:* inverted so that the funicle and micropyle are adjacent.

*Androecium:* the stamens considered as a whole; the male part of the flower.

*Aneuploid:* irregular chromosome number due to loss or gain of single chromosomes, not whole sets.

*Annual:* completing its life cycle within a year.

*Anterior:* facing outwards away from the axis, usually towards the bract.

*Anther:* the part of a stamen which contains pollen.

*Apetalous:* without petals.

*Apical:* at the apex.

*Apocarpous ovary:* with the carpels free from one another.

*Apomictic:* producing seed without fertilization.

*Appressed:* pressed flat and close to an organ.

*Ascending:* curving upwards.

*Assimilates:* products of photosynthesis.

*auct.:* of authors, i.e. has been used but is not valid name.

*Autogamous:* self-pollinating.

*Awn:* bristle-like structure on lemma, etc.

*Axil* of leaf: the angle between the leaf and stem.

*Axillary:* arising in the axil of a leaf or bract.

*Beak* of a fruit: a narrow prolongation.

*Berry:* a fleshy fruit without a hard layer of pericarp around the seeds, usually with several seeds.

*Biennial:* completing its life within two years and flowering in the second year only.

*Bifid:* deeply divided in two.

*Blade* of a leaf: the flat part or lamina.

*Bract:* a leafy structure beneath a flower or group of flowers.

*Bracteole:* a secondary bract, as on the pedicel of a flower.

*Bulb:* a swollen underground bud with fleshy scales and/or leaf bases on a short stem.

*c.:* abbreviation of *circa,* about.

*Calyx:* the sepals considered as a whole.

*Cambium:* a layer of cells dividing to produce phloem externally and xylem internally. See also *cork cambium.*

*Canopy:* leaf cover.

*Capitate:* having a rounded head.

*Capsule:* a dry dehiscent fruit derived from two or more united carpels.

*Carpel:* one of the units composing the gynaecium (or pistil) and containing one or more ovules.

*Caruncle:* a warty or fleshy outgrowth from the surface of a seed, near the micropyle.

*Caryopsis:* a dry one-seeded fruit with the pericarp and testa fused together.

*Cauline leaves:* borne on an aerial stem.

*Cell:* basic unit of plant structure.

*Cellulose:* complex carbohydrate with long unbranched chains of glucose isomer; constituent of plant cell walls.

*Central fusion nucleus:* diploid nucleus formed by fusion of two haploid nuclei of embryo-sac, and fusing with male nucleus to give triploid endosperm nucleus.

*Chlorenchyma:* tissue containing chlorophyll.

*Chlorophyll:* the green pigment of plants, concerned with photosynthesis.

*Chromosome:* thread-like structure in nucleus carrying genetic information.

*Ciliate:* (i) having cilia or flagella, (ii) having a marginal fringe of long hairs.

*Cleistogamous flowers:* not opening, pollination taking place within the closed flower.

*Coleoptile:* bladeless first leaf of grass seedling.

*Collenchyma:* tissue consisting of cells with walls strengthened by layers of cellulose, with the thickening mainly at the corners.

*Commercial variety:* cultivar produced by repeated seeding without special selection of mother seed.

*Common catalogue:* combined national lists of E.E.C. countries.

*Compound leaves:* composed of two or more separate leaflets.

*Connate:* united with another part of the same kind.

convar.: abbreviation of convarietas.

*Convarietas:* taxon smaller than a subspecies, but including several groups, each referred to as a varietas.

*Cork cambium:* a layer of cells which divides to produce cork cells.

*Corm:* a rounded, swollen, fleshy underground stem, outwardly resembling a bulb, but solid.

*Corolla:* the petals considered as a whole.

*Corpus:* central part of apical meristem.

*Cortex:* the tissue between the epidermis (or the piliferous layer of roots) and the endodermis or starch sheath.

*Cotyledon(s):* the first leaf (leaves) of a seed forming part of the embryo.

*Culm:* the aerial stem of a grass or sedge.

*Cultivar:* cultivated variety, named variety in agricultural or horticultural sense; term used in scientific literature to avoid possibility of confusion with botanical variety or varietas.

*Cuneate:* triangular and attached at the point; of leaf bases.

*Cuticle:* impervious layer on surface of epidermis.

*Cutin:* wax-like substance, main constituent of cuticle.

cv.: abbreviation for cultivar.

*Cyme:* inflorescence in which each axis terminates in a flower, with growth continued by side branches.

*Cymose:* in the form of a cyme.

*Cytoplasm:* the protoplasm other than the nucleus.

*DDM:* digestible dry matter.

*Deciduous:* falling off, usually when mature.

*Decumbent:* lying on the ground but ascending at the end.

*Decurrent leaf:* having the base prolonged down the stem often in the form of wings or projections.

*Decussate leaves:* in pairs, opposite, each pair at right angles to the succeeding pair.

*Dehiscent:* opening to shed seeds (or spores).

*Diam:* an abbreviation of diameter.

*Diarch:* having two strands of xylem.

*Diaspore:* structure by which plant is propagated, whatever its morphological nature; used here mainly in sense of 'agricultural seed'.

*Dioecious:* having male and female flowers on different plants.

*Diploid:* with two sets of chromosomes.

*Dissected:* deeply cut into narrow lobes.

*Distichous:* in two opposite vertical ranks.

*DM:* dry matter.

*DODM:* digestible dry organic matter, i.e. DDM less ash.

*Drupe:* a fruit with the outer part fleshy and the seed enclosed in a woody layer of pericarp.

*D-value:* DODM expressed as percentage of DM.

*Ear:* general term for cereal inflorescence.

*Emarginate:* with a shallow notch at the apex.

*Empty glume:* obsolete term for glume.

*Endodermis:* a single layer of cells between the cortex and the vascular-tissue.

*Endosperm:* food-storing tissue formed after fertilization outside the embryo of a seed.

*Ephemeral:* short-lived.

*Epicalyx:* a whorl of sepal-like lobes close to the clayx.

*Epicotyl:* seedling stem above cotyledon node.

*Epidermis:* the outer layer of cells.

*Epigeal:* above ground; in epigeal germination the cotyledons appear above the soil.

*Epipetalous:* attached to the petals.

*Exstipulate:* without stipules.

*Extravaginal:* outside the sheath.

*Extrorse anthers:* opening towards the outside of the flower.

f.: abbreviation for forma.

*F.1:* first filial generation; of cultivars, produced by controlled crossing of two inbred lines and hence uniform and showing hybrid vigour.

*Family:* group of related genera, smaller than an order, larger than a sub-family or tribe.

*Fertilization:* fusion of nuclei in embryo-sac.

*Fibre:* elongated sclerenchyma cell.

*Filament:* the stalk of a stamen.

*Flexuous:* wavy.

*Floret:* (a) small flower, (b) in *Gramineae*, flower plus lemma and palea.

*Flowering glume:* obsolete term for lemma.

*Foliar spray:* spray applied to leaves or growing crop generally.

*Follicle:* a dry several-seeded fruit formed from one carpel and dehiscing along the inner side.

*Forma:* taxon smaller than subvarietas.

*Fortified:* of flour, with added nutrients.

*Fructification:* a general term for a structure bearing spores or containing seeds.

*Fruit:* the ripened gynaecium (pistil) containing the seeds. Some so-called fruits include additional parts such as the succulent receptacle in strawberry.

*Funicle:* the stalk of an ovule.

*Gametes:* male and female haploid cells which fuse to give diploid zygote.

*Gametophyte:* generation producing gametes; alternates with sporophyte.

*Gamopetalous:* having the petals united.

*Gamosepalous:* having the sepals united.

*Genera:* plural of genus.

*Geniculate:* bent abruptly or 'kneed'.

*Genome:* chromosome set of a particular type.

*Genus:* group of related species; the first part of the Latin name of a plant is that of the genus.

*Gibberellic acid:* substance promoting growth in length; originally isolated from the fungus *Gibberella*.

*Glabrous:* not hairy.

*Gland:* structure secreting essential oil or other special product.

*Glandular:* acting as gland, provided with glands.

*Glaucous:* of a pale bluish-green, often somewhat waxy, appearance.

*Glume:* bract of spikelet in *Gramineae*, not subtending a flower.

*Gynaecium:* the female part of a flower including the ovary or ovaries and the style(s) and stigma(s).

*Habit:* the general form of growth of a plant.

*Habitat:* the place in which a plant grows, plus the external factors associated with that place which affect the growth of the plant.

*Halophyte:* a plant which can grow in soil containing appreciable amounts of common salt.

*Haplocorm:* corm formed from single internode of stem.

*Haploid:* with single set of chromosomes.

*Hemiparasite:* obtaining part of its food from a host plant.

*Herbaceous:* not woody; of plants dying down to ground level in winter.

*Hermaphrodite:* having both male and female parts.

*Heterosis:* hybrid vigour.

*Heterozygous:* with homologous chromosomes carrying different genetic information.

*Hexamerous:* floral parts in sixes.

*Hexaploid:* with six sets of chromosomes.

*Hexarch:* with six xylem groups.

*Hilum:* of a seed, the scar left by the stalk (funicle).

*Hirsute:* with rather long, rough hairs.

*Hispid:* bearing stiff hairs.

*Hoary:* covered with very short hairs which produce a whitish appearance.

*Homoeologous:* similar but not identical; of chromosomes, not strictly homologous because belonging to different original genomes.

*Homologous:* corresponding chromosomes of same genome; homologous chromosomes pair at meiosis.

*Homozygous:* carrying the same genetic information on the two members of a pair of homologous chromosomes.

hort.: of gardens; used of names employed in horticulture, but not valid.

*Hypocotyl:* the part of the axis between the root and the cotyledons.

*Hypogeal:* below ground; in hypogeal germination the cotyledons remain below the ground.

*Imbricating:* overlapping; like the tiles on a roof.

*Improvers:* substances added to flour to improve bread-making quality.

*Inbred line:* largely homozygous line produced by continued selfing.

*Indehiscent:* not opening to release seeds or spores.

*Indigenous:* native, not introduced from elsewhere.

*Inflorescence:* a group of flowers on a common axis.

*Infraspecific:* below level of species.

*Infructescence:* inflorescence in fruiting stage.

*Integument:* outer part of ovule; one- or two-layered, growing around nucellus and developing into testa.

*Internode:* the part of the stem between two successive nodes.

*Interspecific:* between species.

*Intraspecific:* within a species.

*Intravaginal:* within the sheath.

*Introrse anthers:* opening towards the centre of the flower.

*in vitro:* in glass; a test carried out using chemical apparatus.

*in vivo:* in the living animal.

*Involucre:* one or more whorls of bracts usually below a compact inflorescence (as in *Compositae*).

*Irregular:* not symmetrical about more than one diameter (zygomorphic).

*LAI:* leaf area index.

*Lamina:* of a leaf; the flat part or blade.

*Land race:* local population of a crop plant selected only by climatic and agricultural conditions.

*Lateral:* at the side.

*Latex:* a milky juice.

*Leaf area index:* ratio of leaf area to area of ground.

*Legume:* a dry fruit derived from a single carpel and splitting along both sutures.

*Lemma:* bract subtending a flower in spikelet of *Gramineae*.

*Lignified:* bearing deposits of the woody material lignin.

*Ligulate:* strap-shaped. Also used to indicate possession of a ligule.

*Ligule:* papery prolongation of leaf-sheath at junction with leaf-blade.

*Local variety:* cultivar derived from mother seed selected by local and agricultural factors, and multiplied for only a few generations under different conditions.

*Loculicidal capsule:* dehiscing down the middle of each compartment (loculus), between the partitions.

*Loculus:* a compartment, as in an ovary.

*Lomentum:* a dry fruit, usually elongated, which breaks transversely into one-seeded portions.

*Lyrate:* pinnatifid, but having a large terminal lobe and smaller laterals.

*Megasporangium:* structure enclosing megaspores; ovule in flowering plant.

*Megaspore:* spore giving rise to female gametophyte.

*Megaspore-mother-cell:* diploid cell which undergoes meiosis to produce megaspores.

*Megasporophyll:* leaf-like structure bearing megasporangia; carpel in flowering plants.

*Mericarp:* a one-seeded portion of a fruit, produced by the ripe fruit breaking into pieces.

*Meristem:* region of actively dividing cells.

*Meristematic tissue:* undifferentiated cells which divide to produce further cells.

*Mesophyll:* parenchyma of leaf.

*Metaphloem:* later differentiating primary phloem.

*Metaxylem:* later differentiating primary xylem.

*Metabolism:* biochemical processes in plant.

*Micro-hair:* small two-celled hair on epidermis of some grass leaves.

*Micropyle:* the minute openiing in the testa of an ovule through which the pollen tube enters; often visible in the resulting seed.

*Microsporangium:* structure enclosing microspores; pollen sac in flowering plants.

*Microspore:* spore which gives rise to male gametophyte; pollen grain in flowering plants.

*Microspore-mother-cell:* Diploid cell which undergoes meiosis to produce microspores.

*Microsporophyll:* leaf-like structure bearing microsporangia; anther in flowering plants.

*Millenium:* one thousand years.

*Monoecious:* having separate male and female flowers on the same plant.

*Monopodial:* formed from a single continuous axis.

*Mucronate:* provided with a minute point.

*Multiple fruit:* fruit formed from more than one flower.

*National list:* list of cultivars accepted in a particular E.E.C. country.

*Node:* the part of a stem from which a leaf arises.

*Nucellus:* a mass of cells within an ovule and surrounding the embryo sac.

*Nut:* a fruit containing a single seed and having a woody pericarp; usually derived from a syncarpous ovary.

*Nutlet:* a small nut, also used for a small nut-like portion of a schizocarpic fruit (*Labiatae* and *Boraginaceae*).

*Ob-:* reversed, e.g. obovate = ovate but attached by the narrow end instead of by the broad end.

*Obtuse:* blunt at the tip.

*Open pollinated:* of a cultivar, multiplied by normal cross-pollination in field, not F.1 hybrid.

*Organ:* a particular plant structure.

*Organelle:* a minute organ present in a cell.

*Order:* a group of related families.

*Orthotropous ovule:* upright, with the micropyle and funicle at opposite ends.

*Outer palea:* obsolete term for lemma.

*Ovary:* the part of the gynaecium containing the ovules.

*Ovoid:* shaped like an egg in three dimensions.

*Ovule:* a structure containing an embryo sac and, after fertilization, developing into a seed.

*P:* probability.

*Paddy:* rice as threshed.

*Palea:* bracteole of *Gramineae.*

*Palisade layer:* specialized layer of mesophyll with closely packed cells elongated at right angles to leaf surface.

*Panicle:* compound raceme.

*Papillose:* covered with papillae (pimple-like projections).

*Pappus:* a ring of hairs or scales at the top of a fruit. A hairy pappus often assists in wind dispersal of the fruit.

*Parenchyma:* a tissue consisting of living cells with uniform thin cellulose walls.

*Pedicel:* a flower stalk.

*Peduncle:* an inflorescence stalk.

*Pelleted:* of seed or other diaspore, enclosed in inert shell to aid precision drilling.

*Pentamerous flower:* having the parts in fives.

*Pentaploid:* with five sets of chromosomes.

*Pentarch:* with five xylem groups.

*Pentosans:* polysaccharides formed from pentose sugars.

*Pepo:* semi-succulent berry derived from inferior ovary of *Cucurbitaceae.*

*Perennial:* persisting for more than two years.

*Perfect flower:* having both male and female parts.

*Perianth:* the part of a flower external to the stamens, including petals and sepals when both present.

*Perianth segment:* a separate leaf of the perianth, usually used when petals and sepals cannot be distinguished.

*Pericarp:* the fruit wall, enclosing the seed(s); derived from the carpel or ovary wall.

*Pericycle:* a layer of non-conducting-tissue one or more cells thick at the periphery of the vascular tissue.

*Periderm:* a protective layer on the outside of parts of some plants, consisting chiefly of cork and cork cambium.

*Perisperm:* a food-storing tissue in some seeds, derived from the nucellus.

*Persistent:* not shed when mature.

*Petaloid:* resembling petals, often coloured.

*Petals:* the inner whorl of the perianth, often brightly coloured.

*Petiole:* the leaf stalk.

*Phloem:* tissue containing sieve tubes in which elaborated foods are translocated.

*Piliferous layer:* epidermis of root, some cells of which are elongated as root-hairs.

*Pistil:* the female part of the flower (gynaecium).

*Pistillate:* having female parts only.

*Pit:* aperture in lignified thickening of a cell wall.

*Pith:* the tissue (usually parenchyma) central to the vascular tissue.

*Placenta:* the place within the ovary at which the ovules are attached.

*Placentation:* the arrangement of the placentae and the ovules.

*Plasmodesma:* cytoplasmic strand passing through cell wall.

*Plumule:* the embryo shoot in a seed.

*Polygamous:* having unisexual and her-maphrodite flowers on the same or on different plants.

*Polypetalous:* having petals free from one another.

*Post-emergence:* applied after appearance of crop above ground.

*Posterior:* of floral parts; facing towards the axis.

*Pre-emergence:* applied before appearance of seedlings above ground.

*Precision drill:* drill capable of sowing individual seeds or other diaspores singly at predetermined intervals.

*Prickle:* a sharply pointed outgrowth of the surface of a plant organ.

*Primary tissue:* tissue formed by differentiation of existing cells, not as a result of cambial activity.

*Probability:* statistical measure of likelihood that results obtained on samples are due merely to chance; probability of 0·01 (= 1%, i.e. such a result would occur by chance only once in a hundred times) accepted as convincing.

*Procambial strand:* strand of undifferentiated cells in young stem which develops into vascular bundle.

*Procumbent stem:* lying loosely on the surface of the ground.

*Prostrate stem:* lying fairly close to the ground.

*Protandrous:* stamens dehiscing before the stigmas are receptive.

*Protogynous:* stigmas receptive before the stamens dehisce.

*Protophloem:* first formed primary phloem.

*Protoplasm:* living material of the plant or animal.

*Protoxylem:* first formed primary xylem.

*Pseudospikelet:* part of inflorescence in bamboos, like spikelet but with additional leafy structures.

*Pubescent:* bearing short, soft hairs.

*Pyxidium:* a capsule which dehisces by a circular slit causing the upper part to form a cap which falls off.

*Quern:* primitive hand mill.

*Raceme:* inflorescence in which growth of single axis is continued, flowers being borne laterally.

*Rachis:* main axis of a grass or cereal spike, on which spikelets are borne.

*Rachilla:* axis of grass spikelet, on which florets are borne.

*Radical:* arising from soil-level.

*Radicle:* the embryo root in the seed.

*Receptacle:* the portion of the axis to which the floral parts are attached. Also used for the enlarged part of the peduncle to which the florets are attached in the *Compositae* and *Dipsacaceae*.

*Recommended list:* selected list including only those cultivars which have given good results in N.I.A.B. trials; usually revised annually.

*Reflexed:* turned sharply backwards or downwards.

*Regular:* symmetrical about more than one diameter (actinomorphic).

*Reniform:* kidney-shaped.

*Reticulate:* forming a network or having a network of surface ridges or markings.

*Rhizomatous:* having rhizomes.

*Rhizome:* underground more or less horizontal stem, usually thickened and perennial.

*Root tuber* or tuberous root: a short swollen root storing food.

*Rotate corolla:* flat, plate-like, not tubular or bell-shaped.

*Rugose:* having a wrinkled surface.

*Runner:* a long prostrate stem rooting at the apex and producing a new plant which later becomes detached from the parent.

*Scarious:* of leaves or bracts, thin dry membranous.

*Schizocarp:* a dry syncarpous fruit breaking up when ripe into one-seeded indehiscent portions.

*Sclerenchyma:* strengthening tissue with lignified cell walls.

*Scur:* short vestigial awn.

*Secondary xylem* and *secondary phloem:* produced from cells derived from the cambium.

sens. lat., *sensu lato:* in the broad sense.

sens. strict., *sensu stricto:* in the narrow sense.

*Sepaloid:* resembling sepals.

*Sepals:* the outer whorl of floral lobes, usually green.

*Septicidal capsule:* dehiscing along the partitions (septa).

*Septum:* a partition, e.g. the wall between neighbouring compartments of an ovary.

*Sessile:* not stalked.

*Shrub:* a short much-branched woody plant.

*Sieve tubes:* tubular cells with perforated transverse walls, present in the phloem and forming a longitudinal system for the translocation of elaborated foods.

*Silicula:* a short broad pod divided into two compartments by a thin septum and dehiscing when mature by the separation of the two valves formed by the pericarp.

*Siliqua:* a fruit similar to a silicula but long and narrow.

*Simple:* of a single piece, not compound.

*Solitary flower* or flower head: borne singly.

*Sp:* an abbreviation of *species* (plural Spp.).

*Spathulate* or *spatulate:* spatula-shaped, like the handle of a spoon.

*Species:* the main unit of plant classification; usually all plants which will cross freely to give fully fertile offspring are regarded as belonging to one species.

*Spicule:* small tooth or spine.

*Spike:* raceme in which florets (or spikelets) are sessile.

*Spikelet:* unit of inflorescence in *Gramineae,* consisting of (usually) two glumes at base and one to many florets borne distichously on rachilla.

*Spine:* a stiff sharply-pointed structure, a modified branch, petiole, stipule or peduncle.

*Sporophyte:* diploid generation, producing spores, alternating with gametophyte generation.

*Staminate:* having male parts (stamens) only.

*Staminode:* a rudimentary or imperfectly developed stamen.

*Stellate:* star-shaped.

*Stigma:* the part of the gynaecium which receives the pollen.

*Stipules:* outgrowths at the base of a leaf.

*Stolon:* more or less horizontal slender stem, usually above ground and then often rooting at the nodes; sometimes used of similar and short-lived stem below ground.

*Stoloniferous:* having stolons.

*Stoma:* aperture in epidermis, bounded by guard-cells.

*Stomata:* plural of stoma.

*Style:* a more or less elongated outgrowth of the gynaecium bearing the stigma.

*Stylopodium:* the enlarged base of a style; as in the *Umbelliferae*.

*Sub-family:* main division of a family; may include several tribes.

subsp., *subspecies:* main division of a species; usually separated by geographical, ecological or genetic barriers.

*Subulate:* awl-shaped, narrowing from the base to a sharp point.

subvar., *subvarietas:* taxon below varietas but above forma.

*Succulent:* soft, thick and juicy.

*Supernumary spikelets:* extra spikelets occurring irregularly on spike.

*Suture:* the line of union of two parts; in fruits dehiscence may take place along a suture.

*Sympodial:* formed by several axes, cymose.

*Syncarpous:* consisting of two or more united carpels.

*Syngamy:* fusion of male and female gametes to form zygote.

*Synonym:* a name not accepted as the valid one.

*Synthetic variety:* one produced by bulking a number of distinct lines.

*Tap-root:* a well-developed vertical main root bearing lateral roots.

*Taxon:* a classificatory group of any size.

*Taxonomy:* the study of systematic classification.

*Tendril:* a slender organ which helps to support a plant by twining around neighbouring stems and other suitable objects. May be a modified stem, leaf or leaflet.

*Tepal:* perianth segment.

*Terminal:* at the end of, terminating.

*Testa:* the outer covering of a seed.

*Tetraploid:* with four sets of chromosomes.

*Tetrarch:* with four xylem groups.

*Thorn:* a modified shoot, leaf or part of a leaf which is woody and sharp-pointed.

*Tiller:* an axillary shoot formed without stem elongation; used of cereals and grasses.

*Tracheid:* an elongated, lignified, water-conducting cell with pointed end walls (see xylem).

*Translocation:* the movement of elaborated substances within the plant.

*Transpiration:* the loss of water from the plant as vapour.

*Transpiration stream:* the upward passage of water and dissolved minerals through the xylem.

*Triarch:* with three xylem groups.

*Tribe:* a group of related genera within a family or sub-family.

*Trifid:* deeply divided into three.

*Trifoliate:* having three leaves or leaflets.

*Triploid:* with three sets of chromosomes.

*Truncate:* appearing to be cut off abruptly.

*Tuber:* a short, swollen underground stem storing food. See also *root tuber.*

*Tunica:* the outer part of a terminal meristem.

*Unilocular:* having one compartment only.

*Unisexual:* of one sex only.

var.: abbreviation of varietas.

*Varietas:* botanical taxon, smaller than subspecies and larger than subvarietas or forma; often anglicized as 'variety', or better as 'botanical variety', but this involves danger of confusion with next entry.

*Variety:* named form of a particular sort of crop plant; may be a clone if vegetatively propagated, a single genotype or a narrow range of genotypes if grown from seed: same as cultivar, which name should be used where any danger of confusion with preceding entry exists.

*Vascular bundles:* strands of food- and water-conducting tissue, containing xylem and phloem.

*Viscid:* sticky.

*Versatile:* with the anther so attached to the filament that it turns freely.

*Vessels:* tubular structures in plants in which water is translocated, part of the xylem.

*Viviparous:* producing living young; used of grasses in which spikelets are replaced by small leafy shoots.

*Whorl:* three or more structures of the same type arising at the same level.

*Xenia:* effect of foreign pollen; used where appearance of seed is altered by cross-pollination with different genotype.

*Xerophyte:* a plant adapted to dry conditions.

*Xylem:* wood; containing vessels and/or tracheids, fibres and xylem parenchyma.

*Zygomorphic:* not symmetrical about more than one diameter, irregular.

*Zygote:* the product of the union of two gametes.

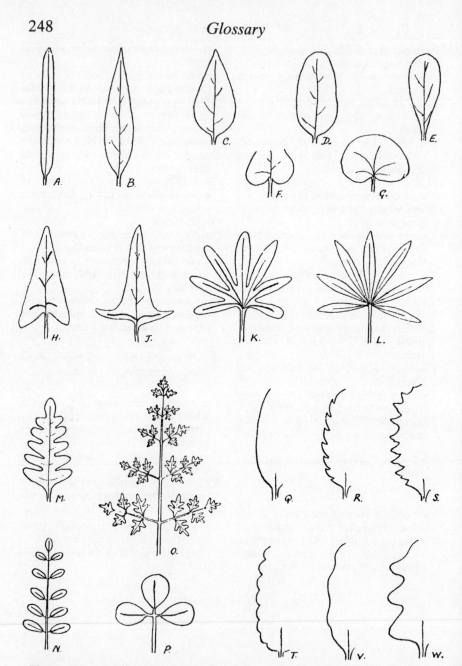

Fig. 95. *Leaf Shapes:* A, linear; B, lanceolate; C, ovate; D, oblong (or oval); E, spathulate; F, cordate base; G, reniform; H, sagittate; J. hastate; K, palmatifid; L, palmate; M, pinnatifid; N, pinnate; O, bipinnate with pinnatifid leaflets; P, trifoliate.

*Leaf Margins:* Q, entire; R, serrate; S, dentate; T, crenate; V, sinuate; W, pinnately lobed.

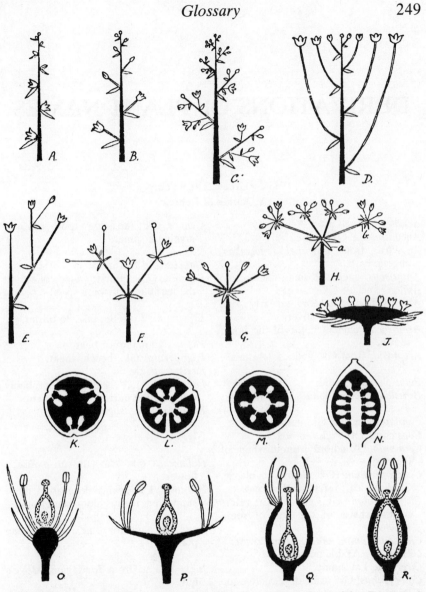

Fig. 96. *Types of Inflorescence:* A, spike; B, raceme; C, panicle; D, corymb (corymbose raceme); E, monochasial cyme; F, dichasial cyme; G, simple umbel; H, compound umbel: *(a)* bract, *(b)* bracteole; J, capitulum.

*Types of Placentation:* K, parietal (T.S.); L, axile (T.S.); M, free central (T.S.); N, free central (L.S.).

*Types of Flower Structure,* showing position of gynaecium (stippled) in relation to the rest of the floral parts (pedicel and receptacle shown in black): O, hypogynous; P, perigynous; Q, perigynous; R, epigynous (receptacle and pericarp fused).

# DERIVATIONS OF PLANT NAMES

Gk.=Greek; Lat.=Latin

## A. Names of Genera

*Achillea:* named after Achilles.

*Anethum:* the Gk. name.

*Angelica:* from Lat. *herba angelica*, angelic herb.

*Anthriscus:* ancient Roman name.

*Apium:* ancient Gk. name.

*Arachis:* from *Arachidna*, an early name of a seed-burying clover.

*Armoracia:* ancient name of the horse radish.

*Artemisia:* from Gk. goddess, Artemis.

*Beta:* classical name.

*Brassica:* ancient name for cabbage.

*Cannabis:* ancient Gk. name.

*Capsicum:* Lat. *capsa*, a box.

*Ceratonia:* Gk. a horn, from the shape of the pod.

*Chaerophyllum:* Gk. *chairo*, to please, *phullon*, leaf; refers to the scent.

*Chenopodium:* Gk., goose foot; refers to the shape of the leaf of some species.

*Chrysanthemum:* Gk., golden flower.

*Cichorium:* Arabic name.

*Cicer:* the Lat. name.

*Colocynthus:* Gk. name of purgative sp.

*Coriandrum:* the classical name.

*Corylus:* the Lat. name.

*Crambe:* Gk., cabbage.

*Crataegus:* Gk., strength; refers to the wood.

*Cucumis:* the Lat. name.

*Cucurbita:* Lat., gourd.

*Cuminum:* the Gk. name.

*Cydonia:* Cydon (now Canea) in Crete.

*Cynara:* Gk., like a dog's tooth; refers to involucre spines.

*Daucus:* old Gk. name.

*Dipsacus:* Gk., thirst; water collects in the leaf bases of some species.

*Erysimum:* from the Gk., to blister.

*Faba:* Lat. name for bean.

*Fagopyrum:* Gk., beech wheat.

*Ficus:* Lat., fig.

*Foeniculum:* Lat. *foenum*, hay; probably refers to the narrow leaf-segments.

*Fragaria:* Lat., fragrance.

*Glycine:* Gk., sweet.

*Helianthus:* Gk. *helios*, the sun, *anthos*, a flower.

*Humulus:* Lat. *humus*, the ground; if not supported, it runs along the ground.

*Isatis:* Gk. name for some dye-producing plant.

*Juglans:* Lat. from *Jovis glans*, Jove's nut.

*Lactuca:* Lat. *lac*, milk; refers to the latex.

*Lavandula:* perhaps from Lat. *lividus*, blueish.

*Lens:* Lat., lentil.

*Lepidium:* Gk., little scale; refers to the fruits.

*Linum:* the Lat. name for flax.

*Lotus:* classical name for some clover species.

*Lupinus:* Lat. *lupus*, wolf; some species were supposed to destroy the soil.

*Lycopersicon:* Gk., wolf peach; may relate to the fact that the fruit was at one time thought to be poisonous.

*Malus:* classical name for an apple tree.

*Medicago:* Medea, where lucerne was supposed to have originated.

*Melilotus:* honey lotus, from the scent.

*Melissa:* from Lat. *mel*, honey.

*Mentha:* ancient Gk. name.

*Mespilus:* the Gk. name.

*Morus:* Lat., mulberry.

*Nicotiana:* after Nicot, a French Ambassador in Portugal, who obtained seeds of tobacco.

*Nasturtium: nasus tortus*, twisted nose; refers to pungent properties.

*Onobrychis:* Gk., asses' food.

*Ornithopus:* Gk., bird foot.

*Pastinaca:* Lat. *pastus*, food.

*Petroselinum:* Lat. *petra*, a rock, *selinon*, parsley.

*Phaseolus:* ancient name of some bean or vetch.

*Pimpinella:* ancient name.

*Pisum:* classical name.

*Plantago:* old Lat. name.

*Polygonum:* Gk., many-kneed; refers to the numerous enlarged nodes.

*Potentilla:* Lat. *potens*, powerful, medicinally.

*Poterium:* from the Gk. for drinking cup; refers to the use of the leaves in cooling drink.

*Prunus:* Lat. name for the plum tree.

*Pyrus:* old Lat. name for the pear.

*Raphanus:* Gk. *raphanos*, appearing quickly; refers to rapid germination.

*Rheum:* old Gk. name from *Rha*, the Volga, from which area rhubarb was obtained.

*Ribes:* From an Arabic plant name.

*Rosmarinus:* Lat., sea dew.

*Rubus:* Lat. *ruber*, red; refers to the colour of the fruits of some species.

*Rumex:* Lat. name.

*Salvia:* from Lat. *salvus*, in good health.

*Sanguisorba:* Lat., blood-stopping.

*Satureja:* the Lat. name.

*Scorzonera:* old French *scorzon,* a serpent; formerly used for snake-bites.

*Sinapis:* old Lat. name for mustard.

*Solanum:* Lat. *solamen*, quieting; refers to its sedative properties.

*Spinacea:* Lat. *spina*; refers to the spiny fruits.

*Symphytum:* Gk., to grow together; refers to the supposed healing properties.

*Tetragonia:* Gk., four-angled fruit.

*Thymus:* the Gk. name.

*Tragopogon:* Gk., goat's beard.

*Trifolium:* Lat., three-lobed leaves.

*Trigonella:* Lat., a little triangle.

*Vicia:* Lat. *vincire*, wind round.

## B. Specific Names

Specific names are usually adjectives the gender of which agrees with that of the generic name. Thus, in association with different generic names, we may have *sativus*, masculine, *sativa*, feminine, and *sativum*, neuter; *arvensis*, masculine or feminine, and *arvense*, neuter; *ruber*, masculine, *rubra*, feminine, and *rubrum*, neuter. Adjectives ending in -*ens*, such as *procumbens*, remain the same in each gender.

Where substantive names (nouns) are used, they are given their own termination, e.g. *Achillea millefolium*. Such names were formerly printed with an initial capital, thus, *Achillea Millefolium, Allium Cepa*, etc. They are often old generic names.

*abyssinicus:* of Ethiopia.
*acaule:* stemless
*acephalus:* without a head.
*acetosus:* acid.
*affinis:* related.
*albus:* white.
*alexandrinus:* of Alexandria.
*ananassus:* resembling the pineapple (*Ananas*).
*andigena:* of the Andes.
*angularis:* angular.
*angustifolius:* narrow-leaved.
*anisum:* from Gk. name.
*annuus:* annual.
*aquaticus:* of water.
*arabicus:* Arabian.
*arietinus:* ram-like (seed shape).
*arvensis:* growing in cultivated fields.
*aureus:* golden.
*avium:* of birds.
*azoricus:* of the Azores.

*basilicum:* Gk., royal.
*biennis:* biennial.
*bonus-henricus:* good Henry.
*botrytis:* clustered, racemose.
*bullatus:* puckered, wrinkled.

*caeruleus:* sky-blue.
*campestris:* of fields or plains.
*caninum:* pertaining to dogs.
*capitatus:* headed.
*cardunculus:* from *carduus*, a thistle.
*carinatus:* keeled.
*carota:* Lat. name.
*caulorapa:* stem turnip.
*cerasus:* cherry-bearing.
*cerefolius:* Lat. form of *Chaerophyllum*.
*chiloensis:* of Chiloe.
*chinensis:* of China.
*cinerariifolius:* leaves like *Cineraria*.
*coccineus:* scarlet.
*commersonii:* (commemorative).
*communis:* growing in colonies; sometimes used in the sense of common or usual.
*corniculatus:* having a little horn.
*creticus:* of Crete.
*crispus:* crisped, curled.
*culinaris:* used in cooking.
*curtilobum:* short lobed.
*cyminum:* Gk. name.

*demissus:* weak, humble.
*domesticus:* domesticated.
*dracunculus:* little dragon.
*dubium:* doubtful.
*dulcis:* sweet.

*esculentus:* edible.

*falcatus:* sickle-shaped.
*ficifolius:* fig-leaved.
*filiformis:* thread-like.
*foenum-graecum:* Greek hay.
*fragiferus:* strawberry bearing.
*frutescens:* becoming shrubby.
*fruticans:* shrubby.

*gaetula:* of the Gaetuli, in N.W. Africa.
*gemmiferus:* bearing buds.
*germanicus:* German.
*gondouinii:* (commemorative).
*graveolens:* strongly-smelling.

*hemicycla:* half circle (pod).
*hispanicus:* of Spain.
*hispidus:* hispid, bristly.
*hortensis:* of gardens.
*hybridus:* hybrid, intermediate.

*idaeus:* Mount Ida, Asia Minor.
*incarnatus:* flesh-coloured.
*indicus:* of the Indies.
*intybus:* Latin name for chicory.

*junceus:* rush-like.
*juzepczukii:* (commemorative).

*laevigatus:* smoothed.
*lanatus:* woolly.
*laurocerasus:* laurel-cherry.
*loganobaccus:* Logan's berry.
*lunatus:* crescent shaped (seed).
*lupulinus:* like a hop.
*lupulus:* *lupus*, wolf, from its tenacious clinging.
*luteus:* yellowish.

*maculatus:* spotted.
*major:* greater.
*maritimus:* maritime, of the sea.
*maximus:* largest.
*medius:* medium, intermediate.
*melongena:* a kind of melon.

*micranthus:* small-flowered.
*microphyllus:* small-leaved.
*millefolius:* thousand-leaved.
*minimus:* smallest.
*minor:* smaller.
*mixtus:* confused.
*monogynus:* having one carpel.
*moschatus:* musky.
*multiflorus:* many-flowered.
*muricatus:* with wall-like markings.
*mutabilis:* changeable.

*napus:* classical name for turnip.
*narbonensis:* of Narbonne, S.W. France.
*niger:* black.
*nipposinicus:* of Japan and China.

*oblongus:* oblong.
*occidentalis:* western.
*officinalis:* officinal, used in medicine.
*oleiferus:* oil-producing.
*oleraceus:* garden vegetable; leaves used in cooking.
*oxyacanthus:* sharp-spined.

*palmatus:* palmate.
*pekinensis:* of Pekin.
*pepo:* Gk. name for gourd.
*perpusillus:* extremely small.
*persicus:* of Persia.
*pimpinellifolius:* with leaves like anise.
*piperita:* from Lat. name of pepper.
*polymorpha:* of many forms.
*porrifolius:* leek-leaved.
*procumbens:* procumbent.

*rapa:* old Lat. name for turnip.
*rapaceus:* turnip-like.

*raphanistrum:* old generic name.
*regius:* royal.
*repens:* creeping.
*reptans:* creeping.
*resupinatus:* upside down.
*rhabarbarum:* rhubarb of the barbarians.
*rhaponticum: rha*, rhubarb, *Pontus*, an area bordering on the Black Sea.
*ruber:* red.
*rusticanus:* of the country.

*sabellicus:* of the Sabines.
*sativus:* cultivated.
*sesquipedalis:* 1½ feet long (pod).
*somniferus:* causing sleep.
*sparsipilus:* sparsely hairy.
*spicatus:* with spikes (of flowers).
*stenotomus:* narrowly cut.
*subterraneus:* underground.
*sylvestris:* growing in woods or forests.

*tabacum:* from native name.
*tataricus:* of Tartary, central Asia.
*tenuis:* slender.
*textilis:* related to textiles.
*tinctorius:* pertaining to dyers.
*tuberosus:* having tubers.

*uliginosus:* growing in marshes.
*uplandicus:* from Uppland, in Sweden.
*usitatissimus:* in most common use.

*veris:* true, genuine.
*viciifolius:* vetch-leaved.
*viniferus:* wine-producing.
*vulgaris:* vulgar, common.
*vulnerarius:* healing wounds.

# UNITS AND CONVERSION FACTORS

Metric units are used throughout this edition, but numerous other units are or have been used in agricultural and botanical literature, and the following notes are given for reference.

Within the metric system standard prefixes are used for fractions and multiples of basic units; the commoner ones are as follows.

*Fractions* (Latin prefixes)
$10^{-1}$ (1/10) deci (d)
$10^{-2}$ (1/100) centi (c)
$10^{-3}$ (1/1 000) milli (m)
$10^{-6}$ (1 millionth) micro ($\mu$, mu)

*Multiples* (Greek prefixes)
10 deca (da)
$10^2$ (100) hecto (h)
$10^3$ (1 000) kilo (k)
$10^6$ (1 million) mega (M)

*Special metric units*
1 litre (l) = 1 cubic decimetre (dm³)
1 cubic centimetre (cc, cm³) = 1 millilitre (ml)
1 are = 100 square metres (m²); hence 1 hectare (ha) = 10 000 m²
1 (metric) quintal = 100 kilograms (kg). Note that quintal originally meant hundredweight (*c.* 50 kg), a similar difference is retained in German where 1 Zentner = 50 kg, 1 Doppelzentner = 100 kg
1 micron ($\mu$, mu) = 1 micrometre ($\mu$m) = 0·001 mm

In some European countries the comma is used instead of the decimal point.

*Approximate conversion factors*

## LENGTH
1 inch (1 in, 1″) = 25·40 mm
1 foot (1 ft, 1′) = 304·8 mm
1 yard (1 yd) = 0·9144 m
1 chain = 22 yards = 20·117 m
1 link = 1/100 chain = 7·92 inches = 201 mm
1 rod (pole, perch) = 5½ yards = 5·03 m
1 line = 1/12 inch = *c.* 2 mm

## AREA
1 square inch (1 sq in, 1 in²) = 645 mm²
1 square foot (1 sq ft, 1 ft²) = 929 cm²
1 square yard (1 sq yd, 1 yd²) = 0·836 m²
1 rod (square measure) = 30¼ yd² = 25·29 m²
40 rods = 1 rood = ¼ acre = 0·101 ha
1 acre = 4 840 yd² = 0·4047 ha

## VOLUME *(British Imperial)*

*Liquid measure*
20 fluid ounces = 1 pint = 0·568 litres
1 gallon = 8 pints = 227¼ in³ = 4·546 litres
(*1 United States gallon = 0·83 Imperial gallons = 3·77 litres*)

*Dry measure, used for corn and seeds*
1 bushel = 4 pecks = 8 gallons = 36·368 litres
1 quarter = 8 bushels = 2·91 hl
(*1 United States bushel = 0·97 Imperial bushels = 36·25 litres*)
(Bushels, being a volume measure, cannot be directly related to weight, but the usual convention was to take the bushel of wheat as 60 lb (*c.* 27 kg), of barley as 56 lb (*c.* 25 kg) and of oats as 50 lb (*c.* 23 kg).)

## MASS (weight)
1 ounce (1 oz) = 28·35 g
1 pound (1 lb) = 16 oz = 0·4536 kg
1 hundredweight (1 cwt) = 112 lb = 50·80 kg
1 ton = 20 cwt = 2 240 lb = 1·016 tonne (t)

## ENERGY AND POWER
1 kilocalorie (Calorie, physiological or medical calorie) = 1 000 calories = 4·187 kilojoules (kJ)

1 standard nutrition unit (1 SNU) = 1 million kilocalories = 4 187 MJ
1 horsepower = *c.* 746 watts (W)
1 watt = 1 J per second

## RATES
(Note that rates can be expressed in three ways: thus kg per ha can also be written as kg/ha or as kg $ha^{-1}$.)
1 oz/yd$^2$ = 33·91 g/m$^2$ = 339 kg/ha
1 lb/acre = 1·121 kg/ha
1 cwt/acre = 125·5 kg/ha; hence 1 fertilizer unit (1/100 cwt) per acre = 1·255 kg/ha
1 Imperial bushel per acre = very roughly, and depending on kind of grain, 60 kg/ha
1 pint/acre = *c.* 1·4 l/ha
Metabolizable energy concentration (MEC) = metabolizable energy (ME) per kg dry matter (DM) = approximately, for most feeds other than oil seeds, D-value/6·3 MJ per kg

# INDEX

Names of plant parts and processes common to most crop plants are not normally indexed except at their first mention or where they are discussed in detail. Names of cultivars, where these are cited as examples, are not indexed; for cultivars the appropriate National List or current NIAB Classified List should be consulted. Latin names in brackets are synonyms.

*Index*